W0050411

Psychological Risks of

Coronary Bypass Surgery

Psychological Risks of

Coronary Bypass Surgery

JUNE B. PIMM
and
JOSEPH R. FEIST

Pimm Consultants
Miami, Florida

With contributions by
Franklin H. Foote
Catherine J. Green
Sally Kolitz
Parry B. Larsen
Theodore Millon
Kathie Roy

PLENUM PRESS • NEW YORK AND LONDON

Library of Congress Cataloging in Publication Data

Pimm, June B., 1927-
Psychological risks of coronary bypass surgery.

Includes bibliographical references and index.
1. Aortocoronary bypass—Complications and sequelae. 2. Aortocoronary bypass—Psychological aspects. 3. Sick—Psychology. 4. Crisis intervention (Psychiatry) I. Feist, Joseph R , 1947- . II. Title. [DNLM: 1. Aortocoronary Bypass—adverse effects. 2. Aortocoronary Bypass—psychology. 3. Coronary Disease. 4. Crisis Intervention. 5. Patient Education—methods. WG 169 P644p]
RD598.P48 1984 617′.412 84-8424

ISBN-13: 978-1-4612-9695-9 **e-ISBN-13: 978-1-4613-2735-6**
DOI: 10.1007/978-1-4613-2735-6

Softcover reprint of the hardcover 1st edition 1984
A Division of Plenum Publishing Corporation
233 Spring Street, New York, N.Y. 10013

To Matthew

Contributors

Joseph R. Feist, Pimm Consultants, 2699 S. Bayshore Drive, Miami, Florida

Franklin H. Foote, Department of Psychology, University of Miami, Coral Gables, Florida

Catherine J. Green, Department of Psychology, University of Miami, Coral Gables, Florida

Sally Kolitz, Department of Psychology, University of Miami, Coral Gables, Florida

Parry B. Larsen, Department of Surgery, University of Miami, Coral Gables, Florida

Theodore Millon, Department of Psychology, University of Miami, Coral Gables, Florida

June B. Pimm, Pimm Consultants, 2699 S. Bayshore Drive, Miami, Florida

Kathie Roy, Cardiovascular and Thoracic Surgery Associates, Suite 207, 1150 N.W. 14th Street, Miami, Florida

Foreword

Heart surgery is still a relatively recent advance in medical technology. The first open-heart procedure was closure of an atrial septal defect in a child at the University of Minnesota Hospital in 1953. This issued in a life-saving advance, the use of which has expanded enormously to include treatment of many areas of cardiac disease. Not unexpectedly, surgical techniques allowed through the use of the heart–lung machine (open-heart surgery) came to be applied in 1967 to the major killer of Americans, namely, coronary artery disease. This operation, known as coronary artery bypass, has become one of the most common surgical operations.

Coronary artery disease, with the possibility of total incapacitation or sudden death from a heart attack, can alter severely the personality of the patient. Corrective surgery can sometimes intensify rather than ameliorate a patient's fears. To the surgeon, occupied by increasing numbers of patients, there is not time enough to give the preoperative attention that might be helpful. Also, the surgeon and cardiologist are limited in their ability to recognize those patients near the breaking point.

The research outlined in these chapters by Drs. Pimm, Feist, and their associates is welcomed by cardiologists and cardiac surgeons. It provides insight into what appears to be reliable recognition of those patients likely to have an adversely affected mental status by coronary bypass surgery and "crisis intervention" to avert this effect and allow the complete benefit of returning the patient to a normal life.

JAMES R. JUDE

Clinical Professor of Surgery
University of Miami School of Medicine
Chairman, Department of Cardiovascular and Pulmonary Disease
Northridge General Hospital
Ft. Lauderdale, Florida

Preface

This book is the result of a personal experience that triggered the idea for a major research study. Approximately five years ago, I encountered a middle-aged couple playing golf. The man appeared strong and muscular and played his first shot with the skill of an experienced athlete. However, as he prepared to hit the ball, his wife turned to me and said fearfully, "I hope he hits a good shot. He has just recovered from heart surgery and is still so depressed; I feel inadequate to help him."

I had just returned from meetings at Harvard University with Gerald Caplan, who had talked of the psychological consequences of normal life crises, and it seemed to me that the experience of coronary bypass surgery, even if successful, could leave someone with an unresolved emotional problem. Caplan proposed "crisis intervention" as a suitable approach for helping individuals in these circumstances, and it occurred to me that it could be helpful for heart surgery patients. I discussed my thoughts with several heart surgeons and was surprised by their immediate interest. Psychological complications, particularly depression, had been a source of worry to many of them.

What followed was the initiation of a long-term research study intended to evaluate the effectiveness of crisis intervention with male coronary bypass patients. Four years later this book describes the project and tells what happened to these patients during the three years after surgery.

Recently someone asked me if our book was completed. Upon hearing my affirmative reply, a stranger asked about its subject and I explained that it dealt with the psychological risks of coronary bypass surgery. To my amazement and delight, this stranger assured me that there was no problem with adverse psychological effects of this procedure if it was carried out at our local hospital. She described her experience when the husband of a close friend underwent surgery at one of the project's two participating institutions. Without knowing that she was describing our study, she glowingly outlined the many ways in which her friends had been helped and supported throughout the experience. For us, none of the enclosed tables of outcome statistics that underline the helpfulness of this approach can equal this unsolicited testimonial from a naive observer. We hope the reader will be equally persuaded that coronary bypass surgery can result in successful outcome, both physically and psychologically.

JUNE B. PIMM

Acknowledgments

It is impossible to know where to begin in acknowledging help in conducting a study of these dimensions. In the beginning our thanks go to the surgeons and cardiologists whose initial interest prompted us to attempt the research. They are James Jude, University of Miami School of Medicine; James Hirschman, Mercy Hospital; Jack Davis, South Miami Hospital; Ernest Traad and John Lister, the Miami Heart Institute; and, with special thanks, Jerry Stolzenburg, the Miami Heart Institute, whose continued interest and support has been indispensable.

We are grateful to the National Heart Association, Miami Chapter, for providing funds for the first year of the research and in particular to James Margolis, chairman of their research committee, who provided suggestions for amending the research design. The Miami Heart Institute made it possible for us to continue the study by providing further funding administered through the office of their capable research director Jeff Raines. Dr. Raines's support for the project, and his belief that the research had merit, has been a key factor in making this book possible.

We thank the patients, the surgeons, and the staffs of the Miami Heart Institute and South Miami Hospital, where the study was carried out. We are particularly indebted to the following surgeons of the Miami Heart Institute: Thomas Gentsch, Parry B. Larsen, Ernest Traad, Malcolm Dorman, and Paul DeWitt. Special thanks go to Dr. Jude, who insisted that we watch him perform a coronary bypass operation before we began our research.

We received invaluable help from University of Miami graduate student assistants Randy Levine, Ron Genellen, and Ken Johnson. We are also indebted to William Kurtines of Florida International University for help with statistical strategies and advice, as well as suggestions for methodology and presentation of results.

Thanks also go to our counselors, Sally Kolitz and Judy Wolfe, who actually met with the patients and their families and provided the supportive counseling which this book is all about. Later, Robin Stillwell provided invaluable insights into the welfare of the patients whom she interviewed 3 years after surgery. Her dedication to the task was remarkable, and her insight into the psychological well-being of each family was formidable. We also gratefully acknowledge the cooperation of wives of our patients who took the time to respond to interviews and fill out questionnaires.

Editorial comments were provided by Robert and Matthew Pimm, whose continued enthusiasm for the writing of the book helped in no small way. Finally, there is no way to express enough thanks to the two women without whom the book would not have been completed: Dolly Lauderdale, our secretary, who typed and revised endless pages, and Agnes Woodward, who volunteered to collect reference material by searching the contents of libraries in two countries. Each in her own way showed enthusiasm, dedication, and loyalty to the project to a truly remarkable degree.

Contents

1

Introduction and Overview

JUNE B. PIMM

Few of us would question that depression often accompanies or follows the experience of coronary artery disease; in fact, this depression has been well documented by numerous research studies (Crisp, DeSouza, & Queenan, 1981; Katon, 1982; Speedling, 1982). Those who experience a heart attack and recover continue to worry about the possibility of another catastrophic event, sudden death, or at the least a reduction in activities and alteration of life-style. The advent of coronary bypass surgery has been viewed by some as meaning an end to the physical and psychological impact of an earlier heart attack or the limitation of activity resulting from living with chronic angina. Since its inception in the 1960s, it has gained in popularity until now over 125,000 coronary bypass operations are performed each year in America; countless others take place in other countries of the industrialized world. Coronary bypass surgery, by alleviating the disabling pain of angina, has held out to heart patients a promise of a normal life, and this implies a life without the psychological handicap of depression.

In spite of these expectations, however, coronary bypass surgery, although highly successful in resolving physical symptoms, has not demonstrated its ability to alleviate some of the incapacitating psychological factors accompanying coronary artery disease. Gloria Hochman in her recent book *Heart Bypass—What Every Patient Must Know* writes,

> Because more than 75% of patients who were previously incapacitated by angina obtain partial or full relief of symptoms after surgery, it would be expected that they would return to work and to productive activity afterward. They do not. (p. 217)

JUNE B. PIMM • Pimm Consultants, 2699 S. Bayshore Drive, Miami, Florida 33133.

This book will describe a project that attempted to help coronary bypass patients deal with the psychological aftermath of the heart bypass experience. The research was originally conceived in 1979 and was triggered by a series of lectures presented by Gerald Caplan, Director of Harvard Medical School's Laboratory of Community Psychiatry. Caplan's persuasive enthusiasm for the concept which he has termed "crisis intervention" sparked an idea which forms the basis for the work which will be described. Briefly, Caplan defines a *crisis* as something which interferes with a person's psychological equilibrium, making it impossible for that individual to continue to function at the same level of efficiency. He argues that these crises can be ameliorated through short-term counseling which he terms *crisis intervention,* and which has as its goal the reinstatement of the individual to his former level of functioning.

Until then there had been only minor interest in the psychological parameters of the surgical experience, with the exception of the landmark works of Janis (1958) and Kimball (1969), which suggested strongly that one's psychological state of mind could have a significant influence on surgical outcome.

Coincidentally, shortly after the author's exposure to Caplan's ideas chance encounters with individuals who had recently undergone coronary bypass surgery suggested that, although these people appeared to have made an excellent physical recovery, they were depressed and unable to function psychologically at their former level. Crisis intervention appeared to have promise as a technique for avoiding this depression. Therefore, a study was planned to investigate its possible usefulness.

Now, 4 years later, the study, which included 104 male coronary bypass patients, has been completed and the results comprise this book. The stated purpose of the project was to ascertain the impact of crisis intervention on the amount and extent of postsurgical depression in males undergoing bypass surgery. In order to do this we took psychological evaluations of patients prior to surgery and at 12 weeks after surgery. Half of the patients participated in crisis intervention counseling and were compared on the amount of their postsurgical depression with a group of patients who did not receive help. Three years after surgery 34 of the original 104 were investigated again for depression to evaluate the possible long-term effect of the earlier crisis intervention.

We also had an interest in the personality characteristics of patients undergoing this type of surgery and the possible interaction between personality characteristics and surgical and psychological outcome. The research, therefore, included a number of psychological measures as well as the depression measure.

Our method involved approaching patients while in hospital during a 2-year period and asking them to participate in research on the emotional concomitants of heart surgery. Our initial intention had been to recruit patients earlier in the surgical process, preferably through the surgeon's office, but this proved not to be feasible for a number of reasons. Because the patients were assigned to treatment and control groups on a random basis, it was difficult to schedule treatment goups from the surgeon's office. Some patients would elect not to follow through with surgery, some would postpone it, others might elect to have surgery done elsewhere or seek a second opinion. For the counselor who needed to schedule her time in order to initiate a certain number of new cases each week, this made a reasonable timetable impossible. An alternative entry to the project would have been the catheterization laboratory, but some of the same problems regarding decisions and booking of surgery would have remained.

The final procedure had the disadvantage of approaching a patient the day before surgery and asking him to participate in a series of time-consuming, paper-and-pencil tests. This had to be fitted into a day already busy with lab tests, presurgical information giving, and visits from medical personnel and family. Also, the crisis intervention counselor (who selected names at random for inclusion in the treatment group) had to schedule her session of anticipatory guidance on the same day. Psychological tests were always completed *before* the anticipatory guidance session in order not to contaminate their results, but the method of picking up the data and returning it to the researchers ran into many snags, and new ways of doing this had to be devised from time to time. There were very few refusals, although some of the patients were unable to complete the entire battery in the time available. Some tired before they had finished. This resulted in incomplete data in a number of cases.

The two participating institutions were the Miami Heart Institute (also the home for the funding during the 1 year of the Miami Heart Association grant and provider of continuing funding the remaining 2 years of the study through its own research funds) and South Miami Hospital. The latter had recently begun to perform coronary bypass surgery and the same group of surgeons used both facilities.

The surgical group who provided the patient population were highly experienced and well recognized in the community and had been performing coronary bypass surgery for several years. Although their medical success rate was high, they had become concerned about the psychological difficulties experienced by some of their patients. This was especially true when these difficulties affected the patient's lack of motivation for returning to work. For the surgeons, returning to work was a meaningful criterion in both psychological and economic terms, and they were anxious to know whether crisis intervention could improve patient motivation in this respect.

Reiterating the problems associated with mounting such a project would be tedious. Research done by clinicians on patients outside a university setting is rarely attempted. One of the greatest difficulties in a pioneering effort is the choice of evaluation instruments. Since there had been very little interest in the effect of psychological intervention on surgical patients, there were few guidelines to help in selection. Fortunately, shortly after the beginning of the project, Theodore Millon of the University of Miami made available to us a scale, the Millon Behavioral Health Inventory. He had developed his scale to assist in predicting both psychological and medical outcome, and our project appeared to provide an excellent opportunity for collaboration. We therefore included it with out measures on this surgical population, and it is described in Chapter 4.

During the 4 years of this research there has been a growing general interest in the relationship between psychological variables and medical and surgical outcome (Gundle, Reeves, Tate, Raft, & McLaurin, 1980; Stanton, Jenkins, Savageau, Harkin, & Aucoin, 1980). What had been a unique idea became more and more commonplace during the time it took to complete the study. However, although there has been an increased interest in the psychological variables associated with heart attacks, and the experience of bypass surgery itself, no studies so far have investigated the helpfulness of a crisis intervention program.

Now that the research is complete, it is exciting to report the results. A problem in this type of research is lack of control over so many possibly influential variables such as the age of the patient, the seriousness of his illness, and the amount or lack of family support. It is gratifying, therefore, that the results appear to point to the effectiveness of the intervention and moreover to its continued impact even at the end of three years. Even more interesting is the interplay between presurgical psychological variables such as recent life events and personality characteristics with medical variables and counseling on postsurgical depression.

Although a great deal of statistical data will be included on the group as a whole, probably one of the most important aspects of this book will be the inclusion of clinical information available on patients. Careful clinical notes were kept by the counselor, and intensive follow-up interviews were conducted both at 12 weeks and at 3 years.

Our interest has been in demonstrating the validity of crisis intervention as a technique for helping coronary bypass patients avoid depression. We did not compare this procedure with alternative intervention strategies such as relaxation, cognitive behavior therapy, information, or hypnosis, nor did we include a control group which received attention only. Had we done so, we should have been able to compare the effectiveness of crisis intervention with alternative treatment styles and could make a more precise statement of its relative effectiveness. This will remain for others to do.

It seemed to us that crisis intervention had a number of advantages over many other treatment strategies. Because it is short-term and not an in-depth approach, it can be used by a variety of helping professions. In its focus on restoring the effectiveness of coping, it fits the prevailing literature on depression appropriately (Seligman, 1972; Weiss, 1982). We feel that it has fulfilled its promise so far as our patients are concerned, and would recommend it for consideration by our readers.

REFERENCES

Crisp, A. H., DeSouza, M., & Queenan, M. *Myocardial infarction and the emotional climate.* Paper presented at the Sixth World Congress of the International College of Psychosomatic Medicine, Montreal, Quebec, Canada, Sept. 13–18, 1981.

Gundle, M. J., Reeves, B. R., Tate, S., Raft, D., & McLaurin, L. P. Psychosocial outcome after coronary artery surgery. *American Journal of Psychiatry,* 1980, *137,* 1591–1594.

Hochman, G. *Heart bypass—What every patient must know.* New York: St. Martin's Press, 1982.

Janis, I. L. *Psychological stress: Psychoanalytic and behavioral studies of surgical patients.* New York: Wiley, 1958.

Katon, W. Depression: Somatic symptoms and medical disorders in primary care. *Comprehensive Psychiatry,* 1982, *23*(3), 274–287.

Kimball, C. P. Psychological responses to the experience of open heart surgery. *American Journal of Psychiatry,* 1969, *126,* 348.

Seligman, M. E. P. Learned helplessness. *Annual Revue of Medicine,* 1972, *23,* 407–412.

Speedling, E. J. *Heart attack: The family response at home and in the hospital.* New York: Tavistock Publications, 1982.

Stanton, B., Jenkins, C. D., Savageau, J. A., Harken, D. E., & Aucoin, R. *Perceived adequacy of patient-education and fears and adjustments after cardiac surgery.* Unpublished manuscript, Boston University School of Medicine and Harvard Medical School, 1980.

Weiss, J. M., Bailey, W. H., Goodman, P. A., Hoffman, L. J., Ambrose, M. J., Salman, S., & Charry, J. M. A model for neurochemical study of depression.

In N.Y. Spiegelstein & A. Levy (Eds), *Behavioral Models and the Analysis of Drug Action. Proceedings of the 27th OHOLO Conference.* Zichron Ya'acov, Israel, 28–31, March, 1982. Amsterdam: Elsevier Scientific Publishing Company, 1982.

I

Coronary Bypass Surgery

Medical, Nursing, and Psychological Factors

2

Coronary Heart Disease

Etiology, Diagnosis, and Surgical Treatment

PARRY B. LARSEN

CAUSE AND PREVENTION OF CORONARY HEART DISEASE

Coronary heart disease (CHD) is the comprehensive term which includes all of the clinical manifestations that result from atherosclerotic[1] narrowing or occlusion of the arteries which supply the heart muscle. The hallmark of clinically manifest atherosclerosis is the progressive deposition of cholesterol as plaques in artery walls. Eventually this leads to secondary changes. Fibrous scar tissue and calcium may accumulate (hence the "hardening"), or the plaques may become necrotic and abscess-like if they outstrip their blood supply. Then hemorrhage into the softened plaque may result in rapid increase in lesion size or the contents may erupt to the surface. If the bloodstream interface of the deposit breaks down, large pieces of cholesterol debris may flake off into the circulation, occluding vessels downstream, and there may be thrombus[2] formation on the rough ulcer surface. When thrombosis occurs in coronary arteries the thrombus usually totally occludes the vessel, producing a myocardial infarction.[3]

Classically, high serum cholesterol, smoking, hypertension, family history, diabetes, and male sex are thought to be the major etiologic

[1]Roughly the equivalent of the more common term, arteriosclerosis or hardening of the arteries.

[2]Loosely but not technically a blood clot.

[3]Death of heart muscle, commonly called a coronary thrombosis or heart attack.

PARRY B. LARSEN • Department of Surgery, University of Miami, Coral Gables, Florida 33124.

factors in CHD (Gordon, Sorlic, & Kannel, 1971). Lowering of serum cholesterol by diet modification and interdiction of smoking have been shown to slow the clinical progression of CHD (Hjermann, Velve Beyre, Holme, & Leren, 1981; Oslo Study Research Group, 1983). Paradoxically, the treatment of low-grade hypertension may even have a detrimental effect on CHD-related mortality, perhaps as a result of the particular medications employed (Multiple Risk Factor Intervention Trial Research Group, 1982; Kaplan, 1983). Treatment of diabetes obviously improves the outlook for the metabolic disorder itself but has no apparent effect on the secondary atherosclerosis. More recently, the presence of Type A personality has been strongly implicated as a causative factor in coronary atherosclerosis although not necessarily of blood vessels elsewhere in the body (Rosenman, Brand, Scholtz, & Friedman, 1976; Haynes & Feinleib, 1982).

CLINICAL PICTURE

Cholesterol deposition may begin fairly early in life (in the 20s and 30s) but is so gradual in its initial stages that symptoms usually do not begin until middle age (late 40s and 50s) or even later. If the cholesterol deposition results in progressive narrowing but not total occlusion of the arterial lumen, a point will be reached at which flow is reduced below the level of heart muscle requirements during exercise. This causes a characteristic constricting pain over the lower sternum known as angina pectoris.[4] Further narrowing of the artery elicits the pain with less and less effort until finally there can be angina at rest. This signifies that even the basal demands of some part of the heart muscle are not being satisfied, an ominous state since any further flow reduction would result in injury or infarction of heart muscle.

At any point in the long process of cholesterol deposition, either before or after angina appears, sudden thrombosis or expansion of a plaque by hemorrhage may totally occlude an artery leading to myocardial infarction. In this situation patients usually experience prolonged chest pain, profuse sweating, and a feeling of prostration. For an unfortunate few patients, the initial symptom of CHD is a catastrophic rhythm disturbance that results in sudden death.

[4]Angina pectoris is classically described by the patient as a constricting or heavy sort of pain as he holds a clenched fist over his sternum. It is precipitated by exercise (or meals) and relieved by rest. The pain may radiate from the mid-sternal area to the shoulder and arm (usually left), jaw, or back. Many variations are known.

Pain is the cardinal symptom of CHD. Rarely does marginal bloodflow during exercise cause angina, nor does infarction of heart muscle cause the prolonged, severe chest pain usually associated with heart attacks. More often an apparent absence of symptoms reflects conscious or unconscious denial by the patient. If the symptoms are not denied outright, they may be ascribed to indigestion and self-medicated with antacids, often for many years. The usual pattern is that patients recognize their symptoms as cardiac in origin and know or even exaggerate the implications as to their mortality. In fact, even the most ardent deniers may withdraw into a depression-like state after an anginal attack without admitting the presence of pain, even to themselves. Whether a patient responds to his anginal attacks with overt anxiety or by withdrawal has major implications for the postoperative period since, as we shall see, unexplained chest pain over the lower sternum is also characteristic of the period following surgery.

DIAGNOSIS AND SELECTION OF PATIENTS FOR SURGERY

The diagnosis of CHD and selection of patients who will require surgery may follow several pathways depending on the mode of clinical presentation. For those who present with angina of effort the first step is the performance of an exercise electrocardiogram[5] (ECG) since the resting ECG is often normal. The patient participates in stepwise increasing exercise on a treadmill while the ECG and blood pressure are monitored. On the basis of the degree of ECG changes and the exercise stage at which the procedure is terminated because of pain the severity of the angina may be classified. This permits fairly accurate prognosis as to longevity, possibility of heart attack, and prediction of the location and severity of atheromatous coronary narrowing (McNeer, Margolis, Lee, Kissla, Peter, Kong, Behar, Wallace, McCants, & Rosati, 1978).

The information from stress tests may be increased if a radioactive tracer (thallium) is injected intravenously just as exercise is terminated. This element is taken up by the heart muscle and will appear on radioactive scans. Areas of heart that pick up the tracer more slowly than others indicate areas of ischemia, that is, poor blood supply.

In many patients the prognosis after these procedures permits a trial of medical treatment with nitrate derivatives (such as isosorbide dinitrate and nitroglycerine), which dilate the coronary vessels, and agents which reduce contractility and pulse rate and hence demand for

[5]Also called a stress test.

oxygen (such as beta-adrenergic blocking agents). Measures directed at risk factor reduction such as cholesterol-reducing diets and cessation of smoking are also very much in order. Even therapy of low-grade hypertension is probably in order although the large multiple risk factor trial will serve as a warning that the drugs must be carefully chosen and monitored (Multiple Risk Factor Intervention Trial Research Group, 1982; Cressman & Gifford, 1983; Pickering, 1983).

Patients who present with angina at rest are often considered to be poor risks for exercise ECG, especially those in whom there has been a recent crescendo increase in frequency, duration, and severity of anginal attacks. Medical therapy is usually given empirically until the acute phase passes and elective stress ECG can be undertaken. Alternatively, if demanding symptoms persist, urgent or emergency cardiac catheterization is recommended.

Patients who present after initially surviving an acute heart attack usually convalesce for 6–8 weeks to permit healing of the infarcted area. However, severe angina returning early after an infarction may precipitate an early need for catheterization. Diagnostic workup including exercise ECG and, if indicated, cardiac catheterization can proceed at lower risk in patients who have thoroughly recovered from their heart attacks. Patients with multiple heart attacks and long-standing heart failure will not be considered here since they are not germane to the study, but they do form a significant proportion of many operative series (Rahimtoola, 1982).

As you have seen, each of the clinical courses may eventuate in cardiac catheterization based on well-defined, intrinsic indications. However, other, more general factors are involved. Younger patients who are in generally good health tend to be moved toward cardiac catheterization and surgery earlier than their opposites. Likewise, patients who are accustomed to high levels of physical activity for work or pleasure are prime candidates. For some individuals an intense psychological response may precipitate or retard catheterization (or surgery). It is not known to what degree these psychological factors determine selection for and the results of surgery.

The cardiac catheterization involves outlining the coronary arteries and left ventricle with an opaque dye while the intensified fluoroscopic image is recorded by high-speed cinephotography. A small catheter is inserted into a peripheral artery (femoral or brachial) and advanced retrograde into the aorta or heart. Generally one or two large boluses of dye are injected into the left ventricle and several injections of very small amounts of dye are made by catheters inserted into the initial portions of the main coronary arteries.

The effectiveness of cardiac pumping is evaluated by the symmetry of ventricular contraction, extent of chamber emptying, and pressure contour accompanying each beat. By examination of the several different cinefluorographic projections during various arterial injections a map of the coronary arteries is obtained and each significant coronary lesion is located on the map and graded as to severity of stenosis.

The entire technical procedure takes 45–120 minutes and has extremely low risk as to life (0.1%) and major complications. The careful explanation of the procedure by the cardiologist and the sedation both contribute to a surprisingly trouble-free time in the catheterization laboratory. Although the injection of local anesthesia, needle and catheter manipulations, and sensation of heat in the head and neck after ventricular dye injection can be moderately uncomfortable or even painful, each step in the process is preceded by explanation and accompanied by reassurance by the staff so that dangerous levels of anxiety are usually avoided. Indeed, the anxiety seems to be displaced to the postcatheterization period in some cases so that the results are impatiently awaited and, at times, unduly emphasized in the patient's own evaluation of his status. Although the cardiologist and surgeon recommend coronary bypass weighing proportionately the clinical course, exercise ECG and scans, and cardiac catheterization, the patient may be overly impressed by a photo or drawing of a 90%-narrowed tube and the analogy (stated or unstated) of what this degree of narrowing might do in the plumbing of his own house.

PREPARATION OF THE PATIENT FOR SURGERY

When the patient presents with angina of effort, surgery is rarely recommended as life-saving unless a severe narrowing of the main left coronary or its equivalent is unexpectedly discovered (Rahimtoola, 1982). Surgery is explained as very successful at relieving angina and improving quality of life, but we have no statistical assurance that heart attacks will be prevented nor life prolonged. Nevertheless, we often hear from families, nurses, and (much later) the patients themselves that surgery was prescribed as a life-saving measure. Perhaps this self-deception is necessary for many patients to accept a procedure in which the heart is deliberately stopped, however small the advertised risk. In addition to presenting a discussion of the benefits, the surgeon informs the patient as to the risk of dying (1.5% to 5%, to be discussed later). Complications are not infrequent and may be severe. It is unnecessarily cruel to describe all of the rare complications, but the patients should know that almost every patient has some complication, that there is a small risk of brain damage, and that surgical correction is not always complete and may even precipitate an acute infarction of heart muscle. The anesthesiologist explains his role and the risks of anesthesia.

Although patient understanding may seem good at the time, key items are often quickly forgotten. Fortunately, the brief appearances of the cardiologist, surgeons, and anesthesiologist are followed by a long session with the nurse educator and usually a representative from the intensive care unit as well. The nurse educator repeats and emphasizes the previous information and responds to questions regarding problem areas from the earlier visits. She conducts preoperative instruction as well. If at all possible, immediate family are included in all of the explanatory sessions or at least subsequently informed of their content.

THE OPERATION

On the day of surgery the patients arrive in the operating room under sedation, which is usually so effective that they do not remember the trip. To ensure this, visits by the family are not permitted once the premedication has been given. However, once the patient is in the operating suite, he may be transiently reawakened by the needles involved in placing cardiovascular monitoring catheters under local anesthesia. Thereafter, general anesthesia is induced.

The chest is opened directly in the midline, splitting the sternum with a saw. On closure with circumferential wires, this provides a much less painful wound than most abdominal or thoracic incisions which cut through thick muscular layers. The pericardium, the serous sac that surrounds the heart, can often be opened in such a way as to avoid entering either of the cavities which surround the lungs while still providing excellent exposure of the entire heart. Large-bore tubes are placed in the right atrium and ascending aorta, and these are connected to a heart–lung machine so that the entire venous return is drained into the oxygenator, aerated by passing bubbles through it, and then pumped into the ascending aorta and out to the whole body.

Once the heart–lung machine takes over the circulation, the heart is stopped and the coronary arteries are bypassed. In most centers there is no way to evaluate the location or degree of coronary artery obstruction in the operating room so that grafts must be placed to the appropriate vessels on the basis of a plan formulated during review of the coronary catheterization preoperatively. In general, all vessels with significant stenosis are bypassed unless they are too small or calcified to permit bypass or the blocked arteries enter areas of total scarring where augmented blood-flow would be fruitless. On the average three arteries are bypassed in each operation, although this may be extended to five or more when branches of the main arteries are individually obstructed.

The usual technique of bypass involves placement of a segment of vein between the aorta and the coronary artery beyond the area of obstruction into a segment of artery that has minimal atherosclerosis. In certain situations the artery which parallels the sternum, the internal mammary artery, is the ideal bypass conduit. The principle is the same. Usually the part of the operation during which the heart–lung machine is employed requires about two to three hours and the whole operation about four hours.

The patient awakens with the same large bore (½ in. or so) tube in his trachea which maintained his respiration during the operation. Often it remains in place until the next morning. Frequently, the patient is paralyzed with medication to permit perfect synchrony with a mechanical respirator. Although this is anxiety provoking and uncomfortable, the combination of preoperative instruction, expert reassurance in the recovery room and intensive care unit, and mild sedation usually permits the patient to sleep most of the time.

IMMEDIATE RESULTS

The overall results of coronary bypass surgery are good. For patients with normal ventricular muscle function as assessed in the catheterization laboratory mortality during the surgical hospitalization is in the range of 1.5% to 3% depending on the extent of atherosclerosis both in the coronary arteries and elsewhere in the body, the patient's general physical condition, and his age (over 70 years of age carries higher risk). Important components of general condition are the presence or absence of significant hypertension, diabetes, and chronic bronchitis and/or pulmonary emphysema—in short, the very diseases that one sees causally linked to atherosclerosis. If ventricular function is poor the immediate mortality rate may rise to 5% or even higher (Rahimtoola, 1982).

Almost all patients experience some complication following the surgery, often of a serious nature. Rhythm distrubances of the heart are the most frequent, ranging from a few extra beats of the ventricles to extremely irregular and rapid rhythms that require immediate treatment. The patient is nearly always very aware of his rhythm disturbance, especially during the period in the intensive care unit when the beep of the monitor reminds him of each heart beat. He also frequently feels a thumping in the chest with each episode of irregularity and receives immediate concerned attention from the nurses. Naturally, it is something less than reassuring to him that the same thumping in his chest can elicit no concern or even notice when it occurs after the monitor has been removed. Of the dysrhythmias which are most dangerous, rapid irregularity of the atria or ventricles may require electric shock for conversion. This is always terrifying, even when there is time to give sedation or anesthesia prior to its administration. Fortunately, most rhythm disturbances are treated with medication. Rhythm irregularities are the most common cause of return to a monitoring area from a more advanced treatment unit. Not surprisingly any regressive movement of this type is accepted as a major set back whether or not this is, in fact, the case. Because new onset of dysrhythmias is fairly common even after the tenth postoperative day, the patient must be warned to report his palpitations and at the same time be reassured that they are probably not important, a fine distinction that may be lost on the anxious patient.

Pain is a cardinal feature of any operation, but, as mentioned previously, it has a double meaning for the post-bypass patient. The pain from the sternal incision may be nearly indistinguishable from angina to some patients. To make matters worse, the pain may subside only to return after 1 week to 10 days as a result of an inflammation of the sac that surrounds the heart called the postcardiotomy syndrome. The associated fever and feeling of prostration and the concern that the doctors seem to have when they listen to a rubbing sound over his heart combine to convince the patient that his condition is very serious indeed. Fortunately, this complication responds rapidly to cortisone-like drugs. The poor localization of anginal pains heightens the uncertainty when the moderate, fleeting pains in unexplainable sites that follow any type of surgery assert themselves in the period following coronary bypass. It may be difficult for the attending surgeon or cardiologist to be certain that the postoperative pain is not recurrent angina, and therefore reassurance may seem half-hearted. To the perceptive patient this may be worse than no reassurance at all.

As with many operations, coronary bypass surgery induces a slightly hypercoagulable state in the blood and this, combined with the circulatory stasis brought on by bed rest, leads to thrombus formation in the leg veins. At the extreme these thrombi can break loose and travel to the lungs, where they can produce serious or even fatal obstruction to the circulation. More commonly, only pain and swelling in the calf are present clincally. Treatment with anticoagulants, elastic hose, and an initial period of bed rest with the legs elevated is required. This, again, is perceived as a major setback since the onset of clinical phlebitis is usually about a week following surgery, when the patient is just beginning to feel good.

A complete discussion of the many complications that can follow coronary bypass would be long and tedious. These few examples are provided because they share the feature that they often occur just as the patient is beginning to feel well and may have serious import. Thus, there may be ample physical reasons for brief periods of depressed mood in the early postoperative period. Other factors will be discussed in the next chapter. However, one group of complications is particularly germane to the main thrust of this book—organic and psychological nervous system abnormalities (Rabiner, Willner, & Fishman, 1975; Slogoff, Girgis, & Keats, 1982).

In the broad sense two factors predispose to brain injury during the course of surgery. First, atherosclerosis tends to be a generalized disease, and one of the sites of most common involvement is in the arteries that feed the brain. At times the atheromas narrow the vessels to a point such that flow may be critically reduced when blood pressure is lower than normal at any time during the operation (especially when the heart–lung machine controls the circulation) or postoperatively. This could, of course, result in a localized stroke. Even when atheromas are not obstructive they may have ulcerated areas from which debris flakes off during the altered flow of cardiopulmonary bypass. This sort of widespread small vessel obstruction produces a more diffuse picture of stroke than does the occlusion of one major vessel. If only tiny amounts of debris break off, the clinical picture may be one of behavioral abnormalities which are difficult to identify positively as organic in origin, or the patient may complain of bizarre visual disturbances which may merge imperceptibly with hallucinations.

The heart–lung machine itself is a source of problems. Great efforts are made to remove the bubbles and platelet aggregates from the blood after it is oxygenated by allowing them to separate in a settling reservoir and microfiltration. At times, however, some particulate matter of microbubbles may pass, producing the same diffuse embolization picture as that seen in atheromatous embolization. Another major source of air embolism is the tubes which are placed in the heart and aorta for connection to the heart–lung machine. At times small quantities of air can enter by this route, and very rarely the air embolism can be massive. This short paragraph only hints at the multiple ways in which the heart–lung machine and its connections can produce neurological complications, but the possibilities for mayhem are obvious.

Psychological stress may accompany complications, but there are other disturbances which are of equal psychological importance. The constantly lighted environment of recovery room and intensive care unit and the frequent awakenings for injections or other treatments lead to total fragmentation of the sleep pattern. The sleep periods themselves are frequently marred by nightmares and semi-awake hallucinations so that the distinction between dream, hallucination, and reality become blurred. Of course, the pattern of interaction with family is totally disrupted as well. Even under the most liberal of regimes, visits are short and communication is frustratingly difficult.

Also, when the patient is awake the menu of sensory experiences is distressingly limited. Reading and television are often either unavailable by doctor's order or frustratingly difficult for the patient in the first few postoperative days. The beep of the monitor and unintelligible distant voices make poor company. The combination of organic dysfunction and sleep and sensory deprivation may precipitate an acute psychosis, which is usually self-limiting and is not highly correlated with later postoperative psychological problems.

POSTOPERATIVE RESULTS

The major benefit from coronary bypass surgery is relief of angina. Symptoms are totally absent in up to 55% of patients in randomized studies and improved in 90%. In selected series which more closely resemble the present study the percentage totally relieved is even higher (Rahimtoola, 1982). Those patients who experience recurrent angina have either incomplete revascularization, graft closure, or progression of atherosclerotic disease. Obviously, the first two causes will tend to appear earlier than the last. The net result is a rate of recurrence of angina of about 4% per year (Rahimtoola, 1982).

In recent years the incidence of late postoperative myocardial infarction appears to be reduced by surgery. More importantly, in the subgroups with severe three-vessel or main left coronary disease there is good evidence of improved long-term survival (Rahimtoola, 1982).

The success of coronary bypass surgery may be assessed in terms other than survival and relief of angina. Fully 30% of patients will have some sexual impairment even when all grafts are patent, and this rises to 50% if all grafts close (Rahimtoola, 1982). Work status is particularly difficult to evaluate (Johnson, Kayer, Pedraza, & Shore, 1982). Coronary bypass surgery, even if successful, tends to accelerate retirement in patients over 60 years old, but this is often elective or based on a physician's nonspecific recommendations to "take it easy." For the younger patients employment rate tends to be higher after successful surgery, and a much larger percentage are able to work without impairment. All too often those who do retire are encouraged to do so by attractive medical retirement plans rather than forced by symptoms.

This chapter and Chapter 3, which discusses the role of intensive nursing involvement in patient education, give a broad overview of the problems involved in the care of the patient who undergoes coronary reconstruction and how they might effect his mental status. It is intended to provide a background against which the psychological interventions that are the subject of this book may be interpreted.

REFERENCES

Cressman, M. D., & Gifford, R. W., Jr. Controversies in hypertension: Mild hypertension, isolated systolic hypertension, and the choice of a step one drug. *Clinical Cardiology,* 1983, *6,* 1–10.

Gordon, T., Sorlic, P., & Kannel, W. B. Coronary heart disease, atherosclerotic brain infarction, intermittent claudication. A multivariate analysis of some factors related to their incidence: Framingham study, 16-year follow-up. Section 27. Washington, D.C.: U.S. Government Printing Office, 1971.

Haynes, S. G., & Feinleib, M. Type A behavior pattern and the incidence of coronary heart disease in the Framingham Heart Study. *Advanced Cardiology,* 1982, *29,* 85–95.

Hjermann, I., Velve Beyre, K., Holme, I., & Leren, P. Effect of diet and smoking intervention on the incidence of coronary heart disease. *Lancet,* 1981, *II,* 1303–1310.

Johnson, W. D., Kayser, K. L., Pedraza, P. M., & Shore, R. T. Employment patterns in males before and after myocardial revascularization surgery. *Circulation,* 1982, *65,* 1086–1093.

Kaplan, N. K. Therapy for mild hypertension. Toward a more balanced view. *Journal of American Medical Association,* 1983, *249,* 365–367.

Leren, P., Helgeland, A., Hjermann, I., & Holme, I. (Oslo Study Research Group). MRFIT and the Oslo Study. *Journal of American Medical Association,* 1983, *249,* 893–894.

McNeer, J. F., Margolis, J. R., Lee, K. L., Kissla, J. A., Peter, R. H., Kong, Y., Behar, V. S., Wallace, A. G., McCants, C. B., & Rosati, R. A. The role of the exercise test in the evaluation of patients for ischemic heart disease. *Circulation,* 1978, *57,* 64–70.

Multiple Risk Factor Intervention Trial Research Group. Multiple risk factor intervention trial. *Journal of American Medical Association,* 1982, *248,* 1465–1477.

Pickering, T. G. Treatment of mild hypertension and the reduction of cardiovascular mortality: The 'of or by' dilemma. *Journal of American Medical Association,* 1983, *249,* 399–400.

Rabiner, C. J., Willner, A. E., & Fishman, J. Psychiatric complications following coronary bypass. *Journal of Nervous Mental Disorders,* 1975, *160,* 342–348.
Rahimtoola, S. H. Coronary bypass surgery for chronic angina—1981: A perspective. *Circulation,* 1982, *65,* 225–241.
Rosenman, R. H., Brand, R. J., Scholtz, R. I., & Freidman, M. Multivariate prediction of coronary heart disease during 8.5 year follow-up in the Western Collaborative Group Study. *American Journal Cardiology,* 1976, *37,* 902–910.
Slogoff, S., Girgis, K. Z., & Keats, A. S. Etiologic factors in neuropsychiatric complications associated with cardiopulmonary bypass. *Anesthesia and Analgesia,* 1982, *61,* 903–911.

3

Coronary Bypass Surgery

The Role of the Patient-Educator

KATHIE ROY

INTRODUCTION

In order to understand the patient-educator's role in caring for the cardiac surgical patient, we must begin with a broad definition of nursing and add to it roles necessitated by the complexity of the care. Henderson defines nursing in this way:

> Nursing is primarily assisting the individual (sick or well) in the performance of those activities contributing to health, or its recovery (or to a peaceful death) that he would perform unaided if he had the necessary strength, will, or knowledge. It is likewise the unique contribution in nursing to help the individual to be independent of such assistance as soon as possible. (Beland, 1965, p. 1)

Nurses caring for most hospitalized patients would expand this definition by accepting responsibility for assessing the emotional needs of patient and family and helping them to deal with this aspect of the illness. Finally, the complexity of modern hospital care requires the strong participation of the nurse in coordinating in-hospital therapy and planning for continuity with external community services when the patient leaves the hospital.

Now that we have outlined the broader aspects of nursing care, we can discuss the special relationship between the nurse as patient-educator and the cardiac surgical patient. Coronary heart disease (CHD) so severe as to require bypass surgery is one of a small set of illnesses which

KATHIE ROY • Cardiovascular and Thoracic Surgery Associates, Suite 207, 1150 N.W. 14th Street, Miami, Florida 33136.

necessitate multiple, high-technology, maximally invasive procedures for diagnosis and treatment. Several doctors and other health care professionals are involved, often for only brief portions of the hospital course. It is difficult for the patient and his family to know the role of each and to know most particularly "who's in charge," since this may shift as the hospitalization progresses. Complex explanations, signing of permits, and diagnostic or surgical procedures follow in rapid-fire sequence. All of this results in a communications crisis. The unique function of the patient-educator is to act as an information channel between all parties. She pays unhurried attention to the details of the patient's problems and questions and obviously knows which of the bewilderingly large array of doctors or other personnel should receive this information. Conversely, if part of the rationale of treatment is poorly understood or forgotten, she is a source of explanation. This builds confidence that the medical team is interested in the patient as a person.

SPECIFIC FUNCTIONS OF THE PATIENT-EDUCATOR

The major specific functions of the patient-educator are (1) detailed description of the usual hospital course to patient and family so that they will know what to expect and what is expected of them; (2) explanation of the disease process and proposed surgery, repeating and clarifying the information given by the doctors; (3) predischarge outlining of steps that the patient can take to speed convalescence and prevent recurrence; and (4) provision of emotional support to patient and family. This last role is particularly important. In our culture the heart is felt to be the seat of the emotions, even the center of life itself. Thus having a disease of so vital an organ is particularly stressful. Because of the large amount of time that the patient-educator spends with the patient early in his hospital course she is uniquely positioned to detect anxiety so great that it might cause dangerously poor coping in the perioperative period. In this case she will suggest to the attending physician the need for consultation with a psychologist or psychiatrist.

This chapter will describe my daily routine as a patient-educator employed by a group of cardiac surgeons in a private practice. Most often I will see the patient at his preoperative work-up in the office; give preoperative instruction in the hospital; help provide continuity of nursing attention as he progresses through recovery room, intensive care unit, telemetry monitoring, progressive care, and discharge; and, finally, see him several times when he returns for follow-up care in the surgeon's office.

BEFORE SURGERY

The most important meeting with the patient and his family, if they are present, is the initial visit. The nurse must create an atmosphere of knowledge, diplomacy, and interest, if a trustful relationship is to be established. This initial meeting will determine the tone for all future encounters.

I find it most helpful and appropriate first to review the patient's chart including the history and physical, lab, and x-ray data, and any consultation notes. Checking the fact sheet to find out the patient's religion, occupation, and locale can also be very helpful. For example, by finding out prior to surgery that a patient lives alone, the discharge planning nurse can be asked to review the patient's chart and meet the patient preoperatively. Thus, by showing the patient that we are interested in his well-being and will make any necessary social service arrangements for discharge at the appropriate time, we can put the patient's mind at ease about what will happen to him in the first few weeks he is at home recuperating.

The purpose of preoperative teaching is to relieve anxiety and stress, dispel misconceptions, and afford an opportunity for the patient and his family to express their fears, concerns, and questions. You may ask yourself, Why should the patient-educator be a nurse and not a doctor? Why should the educator not only be a nurse, but a nurse whose primary responsibility is education of the cardiac patient and his family? First, preoperative teaching takes an average of 30–45 min time, which the doctor and the average floor nurse would have difficulty fitting into their daily routine. Second, the patient-educator provides an element of continuity through the many transfers and single appearances by consultants and therapists that characterize the postoperative period. The nurse teaching cardiovascular patients should be well versed in the unique physical and psychological problems faced by the cardiac patient and his family.

The setting for the teaching should be a room that affords as much privacy as possible to facilitate free flow of communication between the patient and educator. In our hospital setting, cardiac patients very often are disbursed over various hospital areas. There are seldom more than two open-heart procedures per day, easily allowing for individual instruction. In a hospital which does a greater case volume, group teaching might be necessary in order to conserve time. Individual teaching is preferable when possible, because an overanxious individual in a group setting could be quite disruptive during the proceedings. Also, certain individuals are too shy or frightened to ask questions before a group of strangers. Most preoperative teaching is done the day prior to surgery or over several days if there is time.

The preliminary discussion with the patient should elicit what the patient already understands about atherosclerosis, the risk factors of coronary heart disease, the bypass operation itself, and the patient's expectations after his surgery. The aforementioned information should be reviewed with the patient, with clarification of any misconceptions he may have. It is important for the educator to stress that bypass surgery is not a cure for the underlying disease process. When the patient understands his disease, his own positive risk factors, and the fact that surgery is not a cure, later compliance with the discharge information is more probable.

INITIAL PREOPERATIVE PERIOD

The content of the preoperative teaching should include what to expect the day prior to surgery such as ECG, chest x-ray and blood tests, laxative or enema, and the shaving preparation. Very often, the men in particular appear to be disturbed that most of their body hair is to be shaved prior to surgery. This may be seen by the male patient as emasculating. The patient is told of preoperative visits that are to be expected from surgeon and anesthesiologist or, if these have already taken place, further explanation is offered. The patient is instructed to take the sleeping pill when offered even if he feels it is unnecessary, to help him get a good night's sleep, and he is also told that he will receive nothing to eat or drink after midnight the night prior to surgery. Family instructions, including time to see the patient prior to the surgery, location of the surgical waiting area, and the intensive care unit visiting regulations, are reviewed. Preoperative sedation, which includes an antibiotic and sedative, is also mentioned. The patient is told that he will be very relaxed and sleepy when taken to surgery and that he is taken directly to the operating suite where he will see many people in surgical masks and gowns. After being attached to the heart monitor, the peripheral intravenous line is inserted. The patient is told about the insertion of monitoring catheters inserted into various blood vessels and of his coronary catheter. Their purpose and the fact that they are inserted under a local anesthetic is explained. The patient is reassured that the operation is performed under a general anesthetic and that he will hear, feel, and see nothing during this time.

Considerable information on the immediate postoperative period must be given in the preoperative sessions. A patient in the recovery room is in no condition to assimilate new data. After the operation, which can last 4–5 hr, the patient starts responding for short periods of time in the recovery room. Careful explanation is given of the endotracheal tube which is placed through the mouth into the trachea so that respiration may be controlled during surgery. The patient should have a thorough understanding of what the function of this tube is and the fact that he will not be able to speak for as long as it is in place. The patient is reassured that he will never be alone and that even though he is unable to speak with the breathing tube in place, he will be able to shake his head yes or no or even communicate by writing later on. He will feel as though he has a sore throat. Any mucus that collects in his mouth and throat will be suctioned just as it is in the dentist's office. It was my experience as an intensive care unit nurse that the patients not warned about the endotracheal tube were the ones who panicked and needed to be restrained. These patients later recalled feeling as if they were choking and not able to breathe. When a patient is awakening from the effects of his anesthesia, this is not the time to explain a breathing tube. He is too heavily sedated and panic-stricken at this time to assimilate the information. Explanation is given about the nasal oxygen cannula and face tent once the endotracheal tube is removed. The intensive care unit is explained to the patient and his family including the close proximity to other patients, constant noises such as monitoring alarms and bubbling sounds from oxygen and suction devices, and the constant activity of various health team personnel in the room. Having worked as an intensive care unit nurse for 9 years, I have observed that the patient becomes self-centered after surgery. He senses that everything seen and heard pertains only to himself no matter how many other patients are close by. It is therefore very important for personnel not to have medical discussions within earshot of the patient. It is also very important for the staff to explain all procedures as often as necessary in order to alleviate unnecessary fears and anxieties. In our institution most of the patients are on a limited fluid intake for several days after surgery because of fluid retention during the operation. Feelings of dryness and the need for daily weight records are explained. The diet order is usually regular low sodium, and this is started by noon time the day after the operation. Warning about the decreased appetite after surgery is important, for this disturbs many patients and families. The decrease in appetite can be attributed to incisional discomfort, physical inactivity, and decreased taste due to medications. However, it is important to encourage the patient to eat to meet the metabolic requirements of adequate healing.

Since pain is one of the primary concerns of the patient, some time is spent on this subject. It has been my experience that patients undergoing abdominal procedures seem to experience more pain than do open-heart surgery patients. The incision is explained as a clean cut through the sternum or breast bone. Because of the stretching backward of the rib cage and attached muscles, much surgical pain is experienced in the back and shoulders as well. Occasional sticking and stabbing sensations and heaviness over the chest—like a truck sitting on it—are to be anticipated. Patients are encouraged to take the pain medication if needed since this will make it easier to do the required turning from side to side in the bed and coughing and deep breathing exercises every one to two hours. The medication will also make it easier to get out of bed as early as the day following surgery. It is explained to the patient that it is not the moving, deep breathing, and coughing that can cause the most discomfort postoperatively; pulmonary or circulatory complications, if they occur, will cause greater discomfort and require a longer hospital stay. At this time, specific instructions are given in the use of the incentive spirometer and coughing and deep breathing exercises. The patient is then asked to demonstrate his ability. He is encouraged to practice every couple of hours during the day before surgery. The patient is reassured that he will receive pain medication to make this easier.

The patient-educator is in an ideal position to be a troubleshooter at this point. If the patient admits to still smoking or to having stopped recently or if he coughs or expectorates productively, the educator has the responsibility of informing the attending physician and asking for a pulmonary evaluation. Entering into a major operation with compromised lungs can certainly add to the increased morbidity and even mortality a patient may face after this operation.

The intermediate care area or telemetry monitoring area requires explanation. The patient is in a private or semi-private room and has a small battery-operated monitor that transmits his ECG to a master monitor in the nurses' station. He is now assisted in walking about his room and then in the corridors. Family visiting hours return to normal.

It is at this stage in the patient's postoperative course that he may be struck by the intense fatigue and occasional bouts of depression characterized by weeping, anger, and difficulty in concentrating. The profound fatigue is explainable as being due to the major trauma of surgery and the anemia that results from blood loss during and following surgery. This is why patients are placed on iron medication and can expect black stools as a result. The fatigue lessens over a 6–8 week period. Fatigue is probably one of the precipitants for labile feelings of depression after surgery. The anemia and the sleep deprivation caused by being disturbed for medications, testing for vital signs, and breathing exercises can cause occasional bouts of weeping and loss of temper, as well as the poor concentration. Side effects of pain medication and the surgical discomfort itself probably also contribute to these feelings of depression. The patient begins to question whether he should have had the surgery after all. He may feel worse now than he did before the operation. Also, he may start questioning whether he will be able to go back to a productive life after all. If the depression should remain constant or get worse, psychological consultation may be advisable to help the patient cope with his feelings. Because of the atmosphere of trust, the patient-educator may recognize the depression very early and communicate the need for professional therapy to the surgeon or physician. All too often, we focus on the physical aspects of the patient's condition but forget that psychological pain can be far more devastating than any physical discomfort. When a patient is depressed, his physical recovery will take longer. It is the depressed patient who does not care whether he gets pneumonia or circulatory problems. These are the people who very often take longer than 10–14 days in the hospital due to physical problems which are often caused by or aggravated by a negative mental attitude. This in turn adds to increased financial burdens faced by the patient and health care system.

Another side effect after surgery is visual disturbances—blurriness, double vision, stationary objects that appear as if they were moving, flashes through the visual field, or even visual hallucinations. These effects have been studied and many theories given for their occurrence, including medication, sleep deprivation, or the consequences of being on a heart–lung machine. These disturbances usually disappear by the time of hospital discharge or within the first few weeks after the patient is at home.

Before the patient is finally discharged, he is transferred to a progressive care floor. Here he starts showering and is encouraged to ambulate frequently in the hallway. It is at this point that an appointment will be made with the patient and a close family member or friend to discuss discharge.

During the preoperative education session the patient is urged to ask questions, and written literature is left with him reiterating what has been taught. He is also encouraged to watch the hospital video channel, which is programmed with special material that will strengthen his knowledge. The name and office telephone number of the patient-educator are also left with the patient, and he is encouraged to call with questions or if his family have any questions or problems they wish to discuss during the hospital stay.

A written record is then placed in the nursing notes. The record includes content of teaching, special problems or concerns, and details of how the inspirometer was used and if coughing was productive or nonproductive. It is at this point that the physician should be notified about any unusual or significant problems discovered, physical and/or emotional.

The patient-educator uses her unique position to act as a liaison in many areas including planning for discharge or troubleshooting for postoperative respiratory or psychological problems. If she identifies a problem not yet recognized by the attending or consulting physician, she talks with the patient's physician so that a final disposition can be made.

PREPARATION FOR DISCHARGE

The session held to prepare the patient for discharge provides critical information and support. Essentially it is designed to explain and amplify the discharge instructions of the physician and surgeon. Several days prior to discharge, the patient should be notified that he will be going home in the near future and an appointment should be set up with him and a significant family member or friend for a particular day and time. A quiet area should be used for the instruction, which should be done early enough so that a few days are left for follow-up visits prior to the day of discharge for questions or reinforcement. The primary discharge teaching session usually takes 45 min to 1 hr. Whether the discharge teaching is done in a group setting or individually will depend upon the logistics of the situation.

Before a discussion of the discharge information, atherosclerosis and the risk factors of coronary disease are reviewed, with particular attention to any of the positive risk factors the individual patient may have. Our surgeons provide each patient with a small booklet that contains a list of his medications at discharge, an outline summary of his hospital course, and a diagram of his individual bypass procedure, and they review this carefully with him. After the review, the educator informs the patient as to the expected course over the next 4–6 weeks. The aches and pains mentioned in the preoperative teaching sessions are reviewed, as well as what can be done to alleviate them. Measures such as taking Tylenol, using a heating pad on low heat for 20 min 3 or 4 times a day, or a warm bath or shower are recommended. The patient is instructed simply to wash the incisions with soap and water, pat dry, and report any unusual redness or drainage to the surgeon. Lotion containing lanolin may help decrease itching and soften scabs. Women are told that a bra may help relieve pressure on the incision if the breast tissue is heavy. All patients are instructed in the importance of good posture.

The reasons for the profound fatigue are again reviewed with the patient. He is told to take the iron tablets until they are discontinued by the physician, to try to get 6–8 hr of sleep at night, and to take a rest period as needed during the day. However, it is stressed that a rest period does not mean a 20-hr nap and only 4 hr of activity. The patient is told to dress in regular clothes and start getting back to as normal a life as possible. The possibility of "blue days" even at home are mentioned, and what the patient can do to alleviate them if they should occur. For instance, having a good cry may help, or keeping the mind occupied by reading, watching television, having company, or getting out of the house for a while may be helpful. However, if feelings of depression persist, the patient should call the attending physician, who must decide whether the patient should see a psychologist or psychiatrist. A person falling into this category is one who continues to cry a lot, has uncharacteristic changes in behavior such as not caring about physical appearance or remaining aloof, or has unremitting anorexia not related to medication.

Patients are assured that poor appetite will probably improve with increased activity and that any constipation caused by iron can be alleviated with a mild laxative or foods such as prunes or bran.

The patient is questioned about any remaining visual disturbances. If they are still present, the attending physician will often advise him to wait a few more weeks before visiting an ophthalmologist, since it is more than likely that some of the medications such as the sleeping pills and pain relievers may have been responsible for them.

The patient is instructed to notify his attending physician about any unusual pain, fever, or medication problem.

One of the most important subjects and often the most difficult to review is diet. Eating is a social event and most often carries a very pleasant connotation. Diet and eating patterns are also habits which have either been culturally learned or have been self-motivated. Since diets high in saturated fat and cholesterol are felt to be one of the risk factors of coronary heart disease, it is important to instruct the patient in a modified low-fat, low-cholesterol diet, trying to make this as acceptable to the patient and family as possible. The other diet factor that it is important to stress is not adding salt (sodium) at the table. The patient or family member doing the shopping is encouraged to read product labels. If the patient has a special metabolic problem such as diabetes or hypertriglyceridemia, a consultation with a dietician is requested. Helpful pamphlets on diet are also provided to the patient and the family.

Another area that requires special emphasis is exercise. During the first 3–5 weeks at home, the patient is instructed not to participate in activities that require pushing, pulling, excessive bending, or lifting more than 10–15 pounds, since the sternum is still in a healing process. However, since physical inactivity is another risk factor of heart disease, some form of aerobic activity is encouraged on a daily basis. For the first few weeks after surgery walking is the exercise of choice. Walking uses all muscles including the heart but does not require any bending, pushing, pulling, or lifting. The patient is encouraged gradually to increase his walking, using a comfortable pace so that he is walking a total of 2 miles a day at the end of 3–5 weeks at home. He is told this can be accomplished by several walks totaling this distance and is to be done either before eating or at least 1 hr after eating. Walking or other exercise should never be done in extremely hot or cold temperatures and is preferably done on flat surfaces such as indoor malls, hard sand along the beach, or in long corridors. Activities such as swimming, golfing, bike riding, and tennis must first be approved by the physician. Our surgeons do allow our patients to wade in a pool, however, and they may sit outside as long as incisions are not exposed to the sun during the first few weeks at home. The patient is also encouraged to wear his elastic support hose when on his feet if there is still evidence of swelling in the leg where the saphenous vein was harvested for the bypass. Keeping the operative leg elevated when sitting is also encouraged to promote venous return to the heart.

Our patients are told not to drive until their first surgical visit because of the possibility of being thrown into the steering column. Also, reflex time at this early stage of recovery is usually too slow for the individual to drive safely. However, the patient is told that after a week or so, if he feels well enough, he can visit friends, go out for dinner, go to a movie or the like if someone else is driving. Wearing a seat belt or sitting in the back seat of the car is encouraged.

Patients are told that they may have one drink a day with no more than 1½ ounces of alcohol if a mixed drink, or its equivalent in wine or beer. Emphasis is given to making certain that 1½–2 hr is allotted between alcohol and medication administration to avoid untoward reactions.

Sexual relations are permitted as soon as the patient feels physically ready. Quite often it is the spouse who is hesitant to resume sexual relations for fear of hurting the patient. The patient and his partner are reassured that sexual activity will not harm the patient as long as a passive position is used by the patient during the 6–8 weeks it takes the sternum to heal. Reasons for possible impotency such as fear, medication, or physiologic causes such as diabetes or peripheral vascular disease are mentioned and are pursued further if the patient or partner desires.

In our institution the individual cardiologist reviews discharge medications with his patient. The educator stresses the importance of knowing what the medications are and the important facts relating to them. After all, if one puts medicine into one's body, one has the right and the responsibility to know what it is. For example, a patient must understand that because one pill is good, two are not necessarily better. On the other hand, just because he feels good is no reason to discontinue his medication. This is particularly true of antihypertension medication, since high blood pressure has no symptoms *per se* and they may therefore be no warning before a major stroke or heart failure.

Finally, the patient is directed when to make his follow-up visits with the surgeon and cardiologist. At this time all of the information discussed above is left with the patient in written form along with the educator's office telephone number in the event that questions or problems arise once the patient is home. It is at this time that the educator also checks with the discharge planning nurse and makes certain that any arrangements for needed home care follow-up are being followed through.

The patient-educator again charts in the nursing notes all that was discussed and how it was received by the patient and family or friend. I always then attempt to see the patient daily until the day of discharge in case any questions arise or reinforcement of instruction is required. One of the advantages that I have enjoyed in working for surgeons doing heart surgery is that the office follow-up again allows for assessment of compliance with discharge instructions and provides an opportunity for reinforcement of information as needed.

In conclusion, one can see that the nursing role in the care of the cardiac patient is a very important and complex one. Nursing encompasses many roles where the patient and his family is concerned. Nurses act as educators and assessors of physical and emotional problems, providers of physical care and emotional support, and liaison between the patient, family, and the rest of the health care team.

REFERENCE

Beland, I. L. *Clinical nursing: Pathophysiological and psychosocial approaches.* New York: Macmillan, 1965.

4

The Millon Behavioral Health Inventory

Its Utilization in Assessment and Management of the Coronary Bypass Patient

CATHERINE J. GREEN and THEODORE MILLON

Despite impressive advances in medical and surgical treatment, coronary heart disease remains the leading cause of death in the United States today, exacting a terrible toll in both lives and dollars. Coronary bypass surgery has emerged as a widely employed means not only for alleviating angina, but of prolonging life among those with coronary heart disease. The performance of over 100,000 such surgeries annually attests to its utilization and undoubted efficacy. Surgical intervention, however, is not the whole story for it does not take place independently of the person who is experiencing this life-extending treatment.

Several problems may crop up despite excellent surgical results. Although the bypass may provide the foundation for extending and improving the lives of patients, many develop numerous postoperative complications that should not occur either on the basis of their basic disease process or as sequelae of their surgical experience.

This chapter seeks to explore the possible psychosocial aspects of coronary bypass complications, as well as to recommend assessment procedures that may improve postsurgical rehabilitation.

Specifically, we seek to point out a variety of tools and techniques that can be used to ascertain optimal interventions by identifying which

CATHERINE J. GREEN and THEODORE MILLON • Department of Psychology, University of Miami, Coral Gables, Florida 33124.

patients are likely to experience a complicated recovery. We hope to demonstrate that psychological assessment, integrated in a logical and coherent fashion with information on the patient's physical status, can foster these goals. The essential task is that of characterizing certain critical features of the psychological makeup of patients, features which should lead to expeditious ways of modifying negative behavioral influences that can deter or limit the recovery process. It is obvious that individuals who develop coronary heart disease have different life histories and personality traits. Patients are widely distributed along continua of both general emotional and physical health and demonstrate a wide range of strengths and difficulties in dealing with illness. Since physicians must perforce be attentive to myriad aspects of medical evaluation and intervention, they only rarely possess the luxury of time necessary to assess behavioral and emotional issues that may hinder, if not undo, their medical, surgical, and rehabilitative efforts.

Until quite recently, physicians who sought information of this nature were forced to turn to their psychiatric colleagues for time-consuming diagnostic interviews. Similarly, standard psychodiagnostic tests often provided valuable information for the seriously disturbed patients, but these tools were unsuitable and of questionable relevance for the normal but resistant and moderately troubled surgical patient. The majority of psychodiagnostic tests (e.g., MMPI, 16PF) are still employed with medical patients for no reason other than the lack of more relevant tools (Green, 1982).

SEARCH FOR RELEVANT PSYCHOLOGICAL DIMENSIONS IN PHYSICAL HEALTH

It is unfortunate that advances in the physical assessment of medical patients have until quite recently not been matched with comparable advances in psychological assessment.

Faced with the problems and limitations of current psychological tools, the authors, along with their colleague, Robert Meagher, set out to develop a new multidimensional self-report inventory that would be fully relevant to medical and rehabilitation issues. The project began in the early 1970s at a major university medical center. As the literature in this field was reviewed, a confusing and at times almost contradictory array of concepts and findings were found. Closer examination revealed that many of these results were, in fact, complementary, merely differing in their approaches to the patient and the disease process.

One major area of investigation focuses on what may be termed general coping style. These investigators believe that the patient's enduring personality pattern is central to understanding both the development of a disease and how the patient copes with it (Kahana, 1972; Millon, Green, & Meagher, 1982). According to this thesis, enduring psychological tendencies cause individuals to react to stimuli with specific patterns of emotional, cognitive, behavioral, and physiological responses (Lipowski, 1977).

Other researchers have narrowed their attention to the impact of specific psychogenic attitudes rather than more broadly integrated personality styles. These studies typically concentrate on single influences or dimensions; for example, stress is repeatedly found to relate to the incidence of a variety of diseases. More specifically, qualitative studies of chronic stress, such as persistent job tensions or marital problems, have been carried out with particular reference to their impact on heart disease (Friedman & Rosenman, 1974; Jenkins, 1976; Rahe, 1977). Quantitative approaches along the same line of investigation have also been used (Holmes & Masuda, 1974). Many of these studies have focused on the incidence of recent life stress, relating it to the appearance of illness or the exacerbation of a preexistent condition.

Another major area of study may be termed the helplessness–hopelessness constellation. Evidence for the impact of either a premorbid pessimistic attitude or an outlook of future despair on physical disease seems well established. This constellation is in no way illness-specific, however, since there are studies showing its relationship to a wide variety of diseases such as multiple sclerosis, ulcerative colitis, and cancer (Mei-Tal, Meyerowitz, & Engel, 1970; Paull & Hislop, 1974; Schmale, 1972). This depressive pattern has also been investigated in relation to postoperative course; a good surgical outcome appears to be correlated with an acceptance of one's health problems and a positive, future-oriented attitude (Boyd, Yeager, & McMillan, 1973).

Still another significant research perspective pertains to the role of what we have termed *social alienation*. Level of familial and friendship support, both real and perceived, appears to be a significant moderator of the impact of various life stressors (Cobb, 1976; Rabkin & Struening, 1976). All of the stressors seem to be significantly modulated upward or downward by the preoccupations and fears that patients may express about their physical state. Studies of what may be called *somatic anxiety* reflect the general concerns that patients have about their bodies (Lipsitt, 1970; Lucente & Fleek, 1972). As one reflects on these diverse psychogenic components, it becomes apparent that they evidence strong interrelationships. For example, the loss of a spouse means not only dealing with feelings of despair and depression but also dealing with the increased stress of social isolation, the possibility of financial hardship, and a variety of increased responsibilities, each of which serves as an additional source and compounding of stress.

There is a third group of researchers whose primary interest focuses on establishing clear psychosomatic correlates of disease. They study patients with identical physical ailments who can be differentiated in terms of the degree to which psychological factors are central contributors. This realm of investigation is closest to the classic concerns of the psychosomatic clinician; for example, is the patient evidencing an ulcer because of psychic stressors or because of some physiological-nutritional dysfunction? Questions such as these are addressed by physicians who seek to focus their therapeutic attention on the prime source of difficulty; for example, should treatment be geared to psychotherapy, medication, surgery, environmental mangement, or what? Among the major disease syndromes studied in this manner are allergies, gastrointestinal problems, and cardiovascular disorders (Crown & Crown, 1973; Lipowski, 1975; McKegney, Gordon, & Levine, 1970; Pinkerton, 1973).

Still another area addressed by clinicians is the patient's response to illness or treatment and how the patient's psychological makeup affects the course of the illness or the treatment's efficacy. This search for prognostic indices is particularly significant in work with the major life-threatening illnesses and treatment interventions. For example, chronic pain has been extensively investigated with studies of psychological correlates of response to medical or surgical treatment. Other major projects have focused on how patients cope with life-threatening illnesses and whether certain outcomes can be predicted with reference to premorbid personality styles. Identifying which patients will display adverse reactions to surgery or renal dialysis may enable the clinician to institute measures to counteract them and thereby diminish the likelihood of a poor outcome (Abram, 1965; Cohen & Lazarus, 1973).

CONSTRUCTION OF THE MILLON BEHAVIORAL HEALTH INVENTORY

Our search showed that there was no single best instrument or combination of instruments available to meet the varied needs of clinicians interested in problems of physical disease and health. It was the wish of our research group to develop a single instrument that would serve the needs of the health and rehabilitative psychodiagnostician more fully. Brevity, clarity, and ease of administration were added to the goals of elucidating salient and relevant dimensions of functioning. The instrument that was developed over a 4-year span of research was labeled the Millon Behavioral Health Inventory (MBHI). Extensive psychometric data on development, reliability, validity, and norming are available in the manual (Millon, Green, & Meagher, 1982).

In its final form the MBHI consists of 150 self-descriptive true-false items. Table 1 contains a list and brief description of the 20 clinical scales and is divided into four major sections. In the first section are eight scales that comprise the major coping styles; these were derived as normal variants of personality from a theory of personality pathology (Millon, 1969, 1981). The degree to which the patient characteristically exhibits each of the eight styles is expressed in a profile pattern which serves as the basis for interpretation of coping style. The next set of six scales reflects different psychogenic attitudes selected on the basis of their support in the research literature as significant and salient factors that contribute to the precipitation or exacerbation of physical illness—for example, chronic tension and social alienation. The third set of scales was empirically derived by differentiating clients with the same physical syndrome in terms of whether their illness was or was not substantially complicated by social or emotional factors. High scorers on the allergy, gastrointestinal, or cardiovascular scales are those who are most similar to known psychosomatic patients. The fourth set has also been empirically developed. These prognostic indices seek to identify future treatment problems or difficulties that may arise in the course of the patient's illness and rehabilitation.

Machine scoring is available by mail from the publisher, Interpretive Scoring Services Division of National Computer Systems. In line with modern technology, the MBHI can be processed by either computer interactive or teleprocessing methods which synthesize response data into interpretive reports; these clinical assessments are available to all properly qualified medical and psychological professionals. Users with standard terminals and printers can obtain the detailed four-page single-spaced report literally seconds after recording the patient's responses. Interpretive data include a profile of the 20 scale scores and a narrative text that assesses the patient's reaction to his or her illness, manner of relating to health personnel, and likely manner of handling compliance requests and estimates the character and impact of psychological factors upon prognosis and treatment efforts.

In short, what the MBHI was designed to achieve is the expeditious diagnosis of an individual's style of dealing with his illness and with those health care personnel who minister to him. Also available is information

Table 1. Brief Descriptions of High Scores on the 20 Clinical Scales of the MBHI

	Coping styles
Introversive style	Keeps to self, quiet, unemotional, not easily excited, rarely gets socially involved, lacks energy, vague about symptoms, passive about self-care
Inhibited style	Shy, socially ill at ease, avoids close relationships, fears rejection, feels lonely, distrustful, is easily hurt, requires sympathetic support
Cooperative style	Soft hearted, sentimental, reluctant to assert self, submissive with others, lacks initiative, eager to take advice, is compliant, dependent, devalues self-competence
Sociable style	Charming, emotionally expressive, histrionic, talkative, stimulus seeking, attention seeking, unreliable, capricious in affect, easily bored with routine
Confident style	Self-centered, egocentric, narcissistic, acts self-assured, is exploitive, takes others for granted, expects special treatment, is benignly arrogant
Forceful style	Domineering, abrasive, intimidates others, blunt, aggressive, strong willed, assumes leadership role, impatient, and easily angered
Respectful style	Serious minded, efficient, rule conscious, proper and correct in behavior, emotions constrained, self-disciplined, avoids the unpredictable, is orderly, and socially conforming
Sensitive style	Unpredictable, moody, passively aggressive, negativistic, a complainer, guilt ridden, anticipates disappointments, displeased with self and others
	Psychogenic attitudes
Chronic tension	Is under self-imposed pressure, has difficulty relaxing, constantly on the go, impatient
Recent stress	Has experienced significant changes in previous year, life routine has been upset by unanticipated tensions and problems

relevant to future progress. Provided with these data and analyses, the health care clinician should be able to proceed in the best possible fashion, anticipating, as well as managing, difficulties which may arise in the presurgical period and the postsurgical rehabilitation period.

THE USE OF THE MBHI WITH CORONARY BYPASS PATIENTS

As noted earlier, the road to the successful management of the coronary bypass patient is fraught with difficulties. These occur during both the hospitalization period and on return to routine life activities.

Table 1. (*Continued*)

Premorbid pessimism	Is disposed to interpret life as a series of misfortunes, complains about past events and relationships
Future despair	Displays a bleak outlook, anticipates the future as distressing or potentially threatening
Social alienation	Feels isolated, perceives minimal social and family support
Somatic anxiety	Is hypochondriacally concerned with bodily functions, fears pain and illness
	Psychosomatic correlates
Allergic inclination	Is empirically similar to patients evidencing a strong psychological component associated with allergies such as dermatitis and asthma
Gastrointestinal	Is empirically similar to patients evidencing a strong psychological component associated with gastrointestinal disorders such as ulcers or colitis
Cardiovascular tendency	Is empirically similar to patients evidencing a strong psychological component associated with cardiovascular symptoms such as hypertension or angina
	Prognostic indices
Pain treatment responsivity	Is empirically similar to patients who fail to respond successfully to traditional medical treatment regimens for chronic pain syndromes
Life-threat reactivity	Is empirically similar to patients with chronic or progressive life-threatening illnesses—carcinoma, renal failure, and heart disease—who display a more troubled course than is typical
Emotional vulnerability	Is empirically similar to patients who react with severe disorientation, depression, or psychotic episodes following major surgery or other life-dependent treatment programs

An awareness of the individual's concerns and coping style can greatly facilitate a smooth passage in these difficult periods. For example, an individual who is hypochondriacally fearful of his health may have considerable difficulty increasing physical activity, in spite of reassurances from his physician. Such individuals require frequent encouragement and the opportunity to examine and reduce their fears, as well as step-by-step guidance in the rehabilitation program. This avoids the development of superstitious fears and prolongation of the invalid role.

Patient management should be individualized to optimize both treatment response and recovery at each stage of patient management: presurgical, immediately postoperative, and long-term posthospitalization adjustment. The MBHI can serve as a helpful tool in each stage. Sensitive to both the patient's characteristic life-style and more transient concerns, the MBHI directs the clinician's attentions to possible areas of difficulty in the surgical and rehabilitative process.

Several illustrative case reports will be presented to demonstrate the types of patient whose problems of treatment and recovery were minimized through the utilization of MBHI data.

THE MANAGEMENT OF PHILLIP R.

Phillip R. was a 48-year-old attorney with a 9-year history of intermittent angina attacks that appeared during unaccustomed exercise, extreme cold, or emotional stress. Moderately obese, he had been told that his triglycerides were high and was being treated with medication for high blood pressure. Periodic suggestions for diet and weight control were largely ignored, as were recommendations to stop smoking and develop a regular program of cardiovascular conditioning. Hard-driving and extremely successful, he generally paid little attention to his health or the demands he was placing upon his system.

Overdue for his annual check-up, Phillip sought routine attention after a modest increase in anginal symptomatology. The stress test completed at this time indicated the presence of severe coronary artery disease. Immediately hospitalized, he had an angiogram within a few days. Results indicated severe blockage in all three major arteries and surgery was recommended and completed within the week.

At the onset of hospitalization the patient completed the 150-item MBHI; his profile is recorded in Figure 1. An abstract of the computer-generated report notes the following:

> This individual demonstrates conformity of the expectations of others, particularly those in authority. There is a general fear of expressing emotions and losing control, a tendency that often results in tension-produced somatic ailments. There is a strong inclination to deny conflicts and anxieties, and to avoid appearing weak, such as when admitting personal difficulties, especially those of a psychological nature. Due to his reluctance to freely report or discuss ailments, the patient is not likely to be seen as a complainer, even when he has a very deep concern about his health. Anxieties are repressed and isolated from consciousness, often covered up and denied, with a consequent failure to bring symptoms to the attention of health personnel.
>
> This patient experiences family or work responsibilities as persistent and exacting. The demands the patient places upon self, as well as those assumed to meet the desires of others, are felt to be beyond his capacity to fulfill them. Feelings such as these are usually kept beneath the surface, but there may be angry outbursts when these feelings cannot be restrained. Stress of this kind is often associated with tension-related ailments. The past year was viewed as difficult and troubling owing to numerous and unusual life events. Upset by

******MILLON BEHAVIORAL HEALTH INVENTORY***CONFIDENTIAL INFORMATION FOR PROFESSIONAL USE ONLY**

NAME OR NUMBER= Phillip R.
CODED SUMMARY=

SCALES		SCORE RAW	SCORE BR	PROFILE OF BR SCORES (35 · 60 · 75 · 85 · 100)	DIMENSION DESCRIPTION
	1	12	13	XXXX	INTROVERSIVE
	2	6	26	XXXXXXX	INHIBITED
	3	16	23	XXXXXX	COOPERATIVE
PERSNLTY	4	25	50	XXXXXXXXXXXXXXXXXX	SOCIABLE
STYLES	5	27	80	XXXXXXXXXXXXXXXXXXXXXXXXXXXXXXXXXXXXXX	CONFIDENT
	6	16	60	XXXXXXXXXXXXXXXXXXXXXX	FORCEFUL
	7	36	85	XXXXXXXXXXXXXXXXXXXXXXXXXXXXXXXXXXXXXXXXXXXXXX	RESPECTFUL
	8	11	42	XXXXXXXXXXXXXXX	SENSITIVE
	A	17	85	XXXXXXXXXXXXXXXXXXXXXXXXXXXXXXXXXXXXXXXXXXXXXX	CHRONIC TENSION
PSYCHO-	B	9	80	XXXXXXXXXXXXXXXXXXXXXXXXXXXXXXXXXXXXXXX	RECENT STRESS
GENIC	C	4	35	XXXXXXXXXXXXX	PREMORBID PESSIMISM
ATTI-	D	1	18	XXXXX	FUTURE DESPAIR
TUDES	E	9	80	XXXXXXXXXXXXXXXXXXXXXXXXXXXXXXXXXXXXXXX	SOCIAL ALIENATION
	F	9	67	XXXXXXXXXXXXXXXXXXXXXXXXX	SOMATIC ANXIETY
	MM	10	60	XXXXXXXXXXXXXXXXXXXXX	ALLERGIC INCLINATION
PSYCHO-	NN	2	25	XXXXXXXX	GASTROINTESTINAL SUSCEPTBL
SOMATIC	OO	13	58	XXXXXXXXXXXXXXXXXXXXXX	CARDIOVASCULAR TENDENCY
	PP	8	46	XXXXXXXXXXXXXXXXX	PAIN TREATMENT RESPONSE
PROG-	QQ	9	50	XXXXXXXXXXXXXXXXXXX	LIFE-THREAT REACTIVITY
NOSTIC	RR	0	0	X	EMOTIONAL VULNERABILITY

Figure 1. Millon Behavioral Health Inventory for Phillip R.

> these changes and relatively unaccustomed to experiencing them, this patient may now be more prone to major illness than would normally be the case. He reports appreciably less emotional support from family and friends than does the average patient, claiming a lack of caring on their part and an inability to depend upon them in time of need. This sense of isolation and indifference on the part of others is likely to complicate the course and management of an illness.
>
> In spite of these concerns, the patient is inclined to follow rules and look to the doctor for guidance, will probably keep appointments, stick to his early treatment program and prove initially compliant.

The MBHI report alerted Phillip's physician to a number of potential difficulties, as well as suggesting treatment directions following his hospitalization. A problem with coronary heart disease patients is that their symptoms are often of a relatively mild nature and as a consequence surgery may have been experienced as both radical and surprising to them. Add this fact to the dramatic impact of the concept of open-heart surgery. What one often sees in these cases is a strong defensive response to what is experienced as an assault upon one's illusion of invulnerability. Phillip's surgeon not only had to describe the forthcoming intervention but had to deal with the patient's resistance. Although Phillip has a characteristic pattern of denying his difficulties and problems, he demonstrated a heightened alertness to sensations within his body, partly precipitated by the sudden medical attention and the perceived gravity of his circumstances. Receiving detailed information helped Phillip to fully understand the nature and meaning of these sensations and prevented him from ascribing negative and frightening meaning to normal postsurgical sensations. Accustomed to leading a controlled life, Phillip welcomed the opportunity to understand his status and to participate in the planning of his treatment and rehabilitation, even to formulating aspects of his own physical reconditioning and diet.

Postoperative Course and Long-Term Follow-up

As hoped, the patient demonstrated an uneventful postsurgical course, expressing real eagerness to return to work and denying any difficulties of concerns as a function of his surgery, or this demonstration of his mortality. However, he quickly developed an irritated air, giving vent to periodic outbursts when the staff failed to meet his expectations. It proved helpful to understand that this behavior was not directed against the health care professionals but stemmed from Phillip's need to manage and take some control over his life. Having this information, the staff avoided the needless wrangles that so often complicate the daily tasks of clinicians.

In spite of the patient's rapid physical progress, the surgeon had some concerns, particularly because of Phillip's immediate immersion in his work, even while hospitalized, as well as the strain and mutual exasperation demonstrated in many interactions with his wife and teenage son. A brief intervention was instituted that focused on tension-reduction through relaxation training, as well as a cognitive reassessment of Phillip's vocational responsibilities. When he and his wife were interviewed together both were provided with an opportunity to examine fears surrounding the meaning of Phillip's surgery as well as earlier concerns brought to the fore by this life-saving but awesome intervention.

A major difficulty faced with patients like Phillip is their inability to integrate a long-term treatment regimen, despite their general compliance. In fact, one of the first things they will do is convince themselves that they do not have active coronary heart disease and that the circumstances leading them to surgery no longer exist and will not recur. Knowing this, intervention focused on helping the patient to understand and accept the part he played in his illness and to agree to the development of a more adaptive life-style. Although initially reluctant, the patient became aware of some of his unexpressed concerns, acknowledged them, and began to change. Two years later, he is still a hard-working professional but has established some limits on demands, admitting that he always was his own worst enemy. Expecting somewhat less of himself, he shows increased tolerance of family members, conceding their right to autonomy and reducing his sense of responsibility when they do not follow his path.

THE CASE OF JAMES N.

James N. is a 61-year-old former bookkeeper with a 21-year history of cardiovascular problems. The first clinical signs were associated with a sharp increase in blood pressure, which was complicated by his noncompliance with dietary restrictions. Ten years before he had begun to demonstrate signs of coronary artery disease, and these had increased in severity through his fifties. His father had died at 46 of a myocardial infarct, and his younger brother had suffered a severe heart attack 3 years earlier.

Although James had a physician, he did not seek regular medical attention. Living alone since the death of his mother 7 years earlier, he reported his closest relationship to be with a brother who lived in another state. He described himself as somewhat of a loner, particularly since his semiretirement 2 years previously. Hospitalized for surgery, he viewed it as the final event in a long process of decline.

He completed the MBHI upon entry into the hospital for the coronary bypass surgery; the profile is seen in Figure 2. An abstract of his narrative follows:

> The behavior of the patient is typified by a passive-dependency, an amiable manner of compromising in which others are leaned upon for direction and comfort. He may seem weak and express doubts about his ability to manage his affairs, subordinating most personal desires in favor of others and behaving in a self-sacrificing manner in social and family relationships. Despite his mild appearance of genuine agreeableness, he experiences underlying tensions that stems from a mixture of anxious, sad and guilty feel-

******MILLON BEHAVIORAL HEALTH INVENTORY***CONFIDENTIAL INFORMATION FOR PROFESSIONAL USE ONLY*

NAME OR NUMBER= James N.

CODED SUMMARY=

SCALES		SCORE RAW	BR	PROFILE OF BR SCORES (35 / 60 / 75 / 85 / 100)	DIMENSION DESCRIPTION
PERSNLTY STYLES	1	21	68	XXXXXXXXXXXXXXXXXXXXXXXXXXXXX	INTROVERSIVE
	2	16	79	XXXXXXXXXXXXXXXXXXXXXXXXXXXXXXXXXXXXXX	INHIBITED
	3	24	81	XXXXXXXXXXXXXXXXXXXXXXXXXXXXXXXXXXXXXXX	COOPERATIVE
	4	21	22	XXXXX	SOCIABLE
	5	20	38	XXXXXXXXXXXXXXX	CONFIDENT
	6	12	30	XXXXXXXX	FORCEFUL
	7	27	50	XXXXXXXXXXXXXXXXXXX	RESPECTFUL
	8	11	42	XXXXXXXXXXXXXXXX	SENSITIVE
PSYCHO-GENIC ATTITUDES	A	8	32	XXXXXXXXXXX	CHRONIC TENSION
	B	4	40	XXXXXXXXXXXXXXX	RECENT STRESS
	C	7	57	XXXXXXXXXXXXXXXXXXXX	PREMORBID PESSIMISM
	D	11	80	XXXXXXXXXXXXXXXXXXXXXXXXXXXXXXXXXXXXXXXX	FUTURE DESPAIR
	E	11	84	XXXXXXXXXXXXXXXXXXXXXXXXXXXXXXXXXXXXXXXXXXX	SOCIAL ALIENATION
	F	6	46	XXXXXXXXXXXXXXXXX	SOMATIC ANXIETY
PSYCHO-SOMATIC	MM	10	60	XXXXXXXXXXXXXXXXXXXXXX	ALLERGIC INCLINATION
	NN	7	50	XXXXXXXXXXXXXXXXXXX	GASTROINTESTINAL SUSCEPTBL
	OO	15	71	XXXXXXXXXXXXXXXXXXXXXXXXXXXXXX	CARDIOVASCULAR TENDENCY
PROG-NOSTIC	PP	11	59	XXXXXXXXXXXXXXXXXXXXX	PAIN TREATMENT RESPONSE
	QQ	7	40	XXXXXXXXXXXXXXXX	LIFE-THREAT REACTIVITY
	RR	0	0	X	EMOTIONAL VULNERABILITY

Figure 2. Millon Behavioral Health Inventory for James N.

ings. Complaints of weakness and easy fatigability are often expressed when these hidden feelings begin to mount. Under moderate stress, he may readily succumb to exhaustion or apathy, and simple responsibilities may be felt as calling for more energy than he can muster.

The patient may have gone through a period of denying the need for and delaying the seeking of medical treatment since he is inclined both to be embarrassed by and fearful of the implications of physical ailments. Upon initial contact, he may exhibit mixed feelings about doctors and associated health personnel. Although needful of medical reassurance, he has considerable, though often unexpressed, hesitancy about dealing with medical professionals. Moreover, he fears that the doctor will be angry or rejecting if too many complaints are voiced. Once trust is established, however, this attitude may change such that the patient will become overly attached to the doctor, looking for opportunities to receive repeated attention and reassurance. Lacking in confidence and initiative, he may show a greater degree of helplessness than is justified, exhibiting considerable resistance about assuming responsibility for even routine health care. In fact, he may claim to be too confused or too anxious to carry out prescribed treatment regimens without

> close supervision. Disinclined to assume self-responsibility, this patient may become a passive rather than active participant in the treatment process.
>
> This patient feels that his medical condition can only go from bad to worse and that he is not capable of doing anything to prevent this. He feels that he has substantially less emotional support from family and friends than do most people, and states that they do not care about him and that he would be unable to rely on them in times of need. This sense of being alone and unsupported is likely to complicate the management of his illness.
>
> To counter these tendencies and concerns, health care personnel must show a consistently supportive attitude, combined with explicit written instructions given in limited but clearly defined portions over a period of several sessions. This may suffice to establish the trust and self-confidence necessary to activate reasonable compliance on the part of the patient. Helplessness and incompetence appeals should be listened to sympathetically but not acceded to. Although he is disposed to be compliant with those who are trusted, periodic follow-ups are highly advisable in light of his inclination to forget what is disturbing and to depend on others to assume responsibility for the care and monitoring of his health.

This MBHI report confirmed some of the treating physician's fears as well as highlighting other potentially complicating factors. Often the numerous details and issues surrounding coronary bypass surgery would far overshadow the quiet, almost undetectable concerns of this patient. Unwilling to be viewed as troublesome, his characteristic approach is to worry silently, investing his limited energy in negative rumination, rather than focusing on behavior geared toward management of the problem creating the concerns in the first place. The staff's sensitivity to this characteristic pattern directed them to a more encouraging style, one in which physicians and nursing staff alike devoted extra time to drawing James out and encouraging him to share his worries, however minor they might appear. James was grateful for this sensitivity to his needs, feeling comforted in this supportive environment.

Postoperative Course and Long-Term Follow-up

The patient demonstrated an uneventful surgical and immediate postsurgical course, accepting the limits and discomforts of this period well. However, problems appeared when he was asked to shift from the role of completely dependent patient to a more active participatory role in his rehabilitation process. He quickly complained of pain and fatigue, using these sensations as a reason for returning to bed rather than persisting in exercise. When admonished for this attitude, he would become woeful and berate himself for his weakness. Psychological intervention was instituted to facilitate movement toward a more independent stance.

It became apparent that James was very concerned about the nature of his life now that surgery had been completed. Furthermore, he had become quite comfortable in the hospital environment and had little inclination to move into a lonelier, more competitive life circumstance. His therapist focused on development of a larger support network and goal-setting with particular focus on aspects that would provide structured, supportive experiences for the patient. Finally, strategies were implemented to maximize long-term compliance with the treatment regimen. It was feared that James's characteristic response of fatigue would quickly end his much needed exercise program. A buddy system was established to circumvent this difficulty. Two other patients completing postsurgical rehabilitation lived near James, and the three were encouraged to establish a regular pattern of exercise together. Further, James was maintained on an extended, albeit infrequent, outpatient basis both to monitor his compliance and to provide badly needed support. Three years later, James expresses mixed feelings. He sees himself as medically improved and continues to follow the exercise and dieting regimen reasonably well. At the same time, he still feels somewhat isolated and has instituted plans to move to the same city as his brother in an effort to reduce this discomfort.

THE TREATMENT OF GEORGE C.

George C. was an obese, 54-year-old divorced bookkeeper with a long history of multiple medical complaints. He has demonstrated a pattern of frequent illness resulting in seeking of medical attention and repeated evaluation for a wide variety of medical problems. His cardiovascular difficulties are longstanding, with a history of elevated cholesterol levels, hypertension, and angina. Furthermore, his father died of a myocardial infarction at the age of 47 and his older brother suffered a heart attack at 58. In fact, it was his brother's attack that initially brought George to his internist complaining of increased angina and generalized anxiety over his health.

The stress test and other diagnostic evaluations indicated moderate blockage, with good function in spite of this blockage. Surgery was not recommended at that time, but an exercise and diet regimen were established. George began his diet and exercise program; however, within one year he had gained 15 additional pounds, partly in response to the stresses of his second divorce. Hospitalized six months later, he was diagnosed as having a mild heart attack and subsequently reevaluated

******MILLON BEHAVIORAL HEALTH INVENTORY***CONFIDENTIAL INFORMATION FOR PROFESSIONAL USE ONLY**

NAME OR NUMBER= George C.
CODED SUMMARY=

SCALES		SCORE RAW	SCORE BR	PROFILE OF BR SCORES (35 60 75 85 100)	DIMENSION DESCRIPTION
PERSNLTY STYLES	1	12	13		INTROVERSIVE
	2	19	82		INHIBITED
	3	19	52		COOPERATIVE
	4	21	22		SOCIABLE
	5	19	32		CONFIDENT
	6	17	70		FORCEFUL
	7	25	43		RESPECTFUL
	8	21	79		SENSITIVE
PSYCHO-GENIC ATTI-TUDES	A	20	70		CHRONIC TENSION
	B	13	72		RECENT STRESS
	C	30	87		PREMORBID PESSIMISM
	D	24	76		FUTURE DESPAIR
	E	19	76		SOCIAL ALIENATION
	F	27	90		SOMATIC ANXIETY
PSYCHO-SOMATIC	MM	16	81		ALLERGIC INCLINATION
	NN	10	70		GASTROINTESTINAL SUSCEPTBL
	OO	22	85		CARDIOVASCULAR TENDENCY
PROG-NOSTIC	PP	11	59		PAIN TREATMENT RESPONSE
	QQ	12	69		LIFE-THREAT REACTIVITY
	RR	7	80		EMOTIONAL VULNERABILITY

Figure 3. Millon Behavioral Health Inventory for George C.

for coronary bypass surgery. Surgery was recommended and completed at this time.

The results of the MBHI taken by the patient early in his hospitalization are seen in Figure 3.

The following is a brief abstract from the MBHI computer interpretive report:

> This patient is characteristically tense, exhibits an undercurrent of sadness and anger, and is occasionally moody, anxious and irritable, taking a pessimistic and negative outlook on life. Vigilant about what others think, he is watchful for fear of being hurt. There is an inclination to react to events in a somewhat unpredictable manner, with anger and disappointment expressed at one time, usually followed with apologies for being so emotional the next. This emotionality and angry vigilance is both physically and psychologically upsetting and may dispose him to psychosomatic discomforts and ailments. In regard to illness he may be almost exhibitionistic when describ-

ing his symptoms, complaining at great length about an everwidening variety of discomforts, appearing at times to enjoy the role of being ill.

Characteristically negative and inclined to view life as difficult, the patient appears even more negative and distressed than usual, seeing the past as full of misfortunes about which nothing can be done. These attitudes are likely to complicate the course of an illness and its treatment. Further, he views the future as especially problematic. Anticipating that the future will not be productive, but complicated by personal difficulties and worsening medical problems, the patient feels unable to cope well or even to make an effort to do so. The patient reports far less emotional support from friends and family than is typical. In addition to characteristic emotional complaints, this patient is deeply preoccupied with physical health concerns. Anxious over even minor discomforts, he may become very upset by any change in symptoms, which are invariably assumed to be for the worse. Health care personnel, alert to these excessive concerns, should allay anxiety, but retain focus on the real medical issues. If faced with major surgery, there is a chance, albeit small, that this patient will briefly display depression or disorientation.

Patients such as these tend to be as erratic in their relations with doctors as they are with others, alternately engaging and distancing to the exasperation of professionals. He may be inclined to collect doctors and medications, shopping about, rarely satisfied with the results of any treatment or with expertise of all professionals, combining a variety of treatments and medications without consultation or supervision. Sensitive to negative suggestions or exasperation on the part of health professionals, he will become easily upset or angered. It is likely that his cardiovascular complaints are greatly affected by his characteristic hypersensitivity to stress.

The MBHI report clearly indicated that George's health preoccupations were an integral part of a longstanding pattern of cranky unhappiness, and that seeking to change this pattern significantly was unrealistic. Efforts were directed rather to minimize the impact of this characteristic stance upon both the patient and his level of functioning.

When first hospitalized, George's anxiety led him frequently to question his doctors and nurses about any number of minor concerns. The instinct of the clinician is to provide detailed information to patients in response to these inquiries. However, what would occur at that point is that the patient would distort or misconstrue what was said, increasing his anxiety as well as his questioning. When an individual such as this patient is asking the doctor to tell him what will happen, what he actually is requesting is reassurance that everything will be all right. Sensitized to this possibility, the physician provided George with information about the nature of his illness and the surgical procedure while concurrently providing strong statements of the support and confidence of the attending physician. This met the patient's need to trust and, however briefly, comply with the demands he faced.

Postoperative Course and Long-Term Follow-up

The patient's surgery was successfully completed although he was briefly disoriented in surgical intensive care. All health care personnel had been encouraged to be particularly patient in allaying George's concerns. Because of his depression and life stresses, a psychologist was called in for both presurgical counseling and postsurgical follow-up. Anxious and fearful, George required encouragement to become active following his surgery. Provided with a forum in which to examine his fears, the patient was able to complete his postsurgical recovery successfully, albeit with checkered progress. The counseling process focused on long-term compliance. George continued to see the psychologist well into his outpatient rehabilitation program. Anticipating that the return to work would be difficult, George and his therapist developed strategies for his return, detailing his work schedule, a means of maintaining his new fitness regimen, and how and when he might share information regarding his medical status with co-workers. Three and one-half years later, George remains active and employed and continues his diet and exercise program, with occasional lapses. Although he still expresses concerns and difficulties with interpersonal relationships, he senses that he is in command of his life, particularly his physical well-being.

SUMMARY

The MBHI is the first general-purpose instrument of a psychodiagnostic nature designed for use in a wide range of medical and rehabilitation settings. Relevant and brief, it may prove to be a practical tool for obtaining information on both coping styles and specific areas of psychological concern. The instrument has been developed with physically ill patients in mind, and each of the 20 clinical scales can be evaluated on its own or interpreted in conjunction with other test of biographical data. It can serve as a valuable adjunct in the treatment of the surgical patient.

REFERENCES

Abram, H. S. *Adaptation to open heart surgery: A psychiatric study of response to the threat of death.* Paper presented at the meeting of the American Psychiatric Association, New York, May 1965.

Boyd, I., Yeager, M., & McMillan, M. Personality styles in the postoperative course. *Psychosomatic Medicine,* 1973, *35,* 23–40.

Cobb, S. Presidential address—1976: Social support as a moderator of life stress. *Psychosomatic Medicine,* 1976, *38,* 300–314.

Cohen, F., & Lazarus, R. S. Active coping processes, coping dispositions and recovery from surgery. *Psychosomatic Medicine,* 1973, *35,* 375–391.

Crown, S., & Crown, J. M. Personality in early rheumatoid disease. *Journal of Psychosomatic Research,* 1973, *17,* 189–196.

Friedman, M., & Rosenman, R. H. *Type A behavior and your heart.* New York: Knopf, 1974.

Green, C. Psychological assessment in medical settings. In T. Millon, C. Green, & R. Meagher (Eds.), *Handbook of clinical health psychology.* New York: Plenum Press, 1982.

Holmes, T. H., & Masuda, M. Life change and illness susceptibility. In B. S. Dohrenwend & B. P. Dohrenwend (Eds.), *Life events: Their nature and effects.* New York: Wiley, 1974.

Jenkins, C. D. Psychologic and social risk factors for coronary disease. *New England Journal of Medicine,* 1976, *294,* 987–994.

Kahana, R. J. Studies in medical psychology: A brief survey. *Psychiatry in Medicine,* 1972, *3,* 1–22.

Lipowski, Z. J. Psychophysiological cardiovascular disorders. In A. M. Freedman, H. I. Kaplan, & B. J. Sadock (Eds.), *Comprehensive textbook of psychiatry* (Vol. 2, 2nd ed.). Baltimore: Williams & Wilkins, 1975.

Lipowski, Z. J. Psychosomatic medicine in the seventies: An overview. *American Journal of Psychiatry,* 1977, *134*(3), 233–234.

Lipsitt, D. R. Medical and psychological characteristics of "crocks." *Psychiatry in Medicine,* 1970, *1,* 15–25.

Lucente, F. E., & Fleek, S. A study of hospitalization anxiety in 408 medical and surgical patients. *Psychosomatic Medicine,* 1972, *34,* 304–312.

McKegney, F. P., Gordon, R. O., & Levine, S. M. A psychosomatic comparison of patients with ulcerative colitis and Crohn's disease. *Psychosomatic Medicine,* 1970, *32,* 153–166.

Mei-Tal, V., Meyerowitz, S., & Engel, G. L. The role of psychological process in a somatic disorder: Multiple sclerosis. *Psychosomatic Medicine,* 1970, *32,* 67–86.

Millon, T. *Modern psychopathology.* Philadelphia: Saunders, 1969.

Millon, T. *Disorders of personality. DSM III, Axis II.* New York: Wiley, 1981.

Millon, T., Green, C. J., & Meagher, R. B. *Millon Behavioral Health Inventory Manual.* Minneapolis: National Computer Systems, 1982.

Paull, A., & Hislop, I. G. Etiologic factors in ulcerative colitis: Birth, death, and symbolic equivalents. *International Journal of Psychiatry in Medicine,* 1974, *5,* 57–64.

Pinkerton, P. Editorial: The enigma of asthma. *Psychosomatic Medicine,* 1973, *35,* 461–463.

Rabkin, J. G., & Struening, E. L. Life events, stress and illness. *Science,* 1976, *194,* 1013–1020.

Rahe, R. H. Epidemiological studies of life change and illness. In Z. J. Lipowski, D. R. Lipsitt, & P. C. Whybrow (Eds.), *Psychosomatic medicine*. New York: Oxford University Press, 1977.

Schmale, A. H. Giving up as a final common pathway to changes in health. In Z. J. Lipowski (Ed.), *Psychosocial aspects of physical illness*. Basel: Karger, 1972.

II

Techniques for Working with the Coronary Bypass Patient

5

Crisis Intervention and Coronary Bypass Surgery

SALLY KOLITZ and JUNE B. PIMM

INTRODUCTION TO CRISIS INTERVENTION

Crisis intervention is not new. As a therapeutic technique, it has an unusual history stemming from the aftermath of the 1943 Coconut Grove nightclub fire which trapped 425 people. Eric Lindemann, a Harvard psychiatrist, in trying to help survivors deal with their grief, formalized the first mental health approach for dealing with the crisis of loss.

Lindemann was the first to suggest that crises result in the breaking down of coping strategies and that the function of a mental health counselor is to help a person in crisis to reestablish his basic coping style. He also suggested that crisis resolution consists of a series of steps which can be identified and which follow a predictable pattern. His suggestions, published in the *American Journal of Psychiatry* in 1944 under the title, "Symptomatology and Management of Acute Grief," introduced the notion of crisis intervention to the mental health field.

Ten years later, Caplan, in his book *Principles of Preventive Psychiatry* (1964), conceptualized a more general approach suitable to a variety of crisis situations. Since 1954, at the Harvard Family Guidance Center, Caplan had been concerned with the effects of crises. Initially interested in the effect of the birth of a premature infant on young families, over

SALLY KOLITZ • Department of Psychology, University of Miami, Coral Gables, Florida 33124. JUNE B. PIMM • Pimm Consultants, 2699 S. Bayshore Drive, Miami, Florida 33133.

the next 20 years he carefully constructed a conceptual model which he later termed *crisis intervention*.

For Caplan each individual is a constantly developing and ever adapting organism striving to achieve a state of equilibrium or homeostasis. During everyone's lifetime, developmental tasks are encountered which require adaptation. As these are incidents which occur frequently, everyone gradually acquires coping styles or strategies with which to deal with these new demands. Caplan also theorizes that coping strategies are highly personalized and can either be mentally healthy or maladaptive in style. Thus there can be extensive individual differences in coping from one individual to the next.

Crises, according to Caplan, are transition points which require an individual to muster his coping strategies and deal with the emotional distress associated with being "out of equilibrium." He feels that at these transition points persons have the opportunity to move closer or further away from a mentally healthy style of adaptation. Moreover, Caplan observed that during a crisis individuals had a heightened susceptibility to intervention and that mental health intervention appeared to be of more significance at those times.

Although crises can be due to either internal or external changes that necessitate adaptation such as the internal crisis of an illness or the external crisis of divorce, an individual's coping responses are expected to follow a predictable pattern. For example, during the initial stages of a crisis the person's usual pattern of functioning becomes disorganized; he feels anxious and his thought processes are often confused and ineffective. Because the individual finds his traditional coping strategies unsuccessful, he experiences feelings of frustration and helplessness.

Caplan postulated a time frame for a crisis, which he felt lasted from 4–6 weeks. At the end of that time he observed that the emotional distress usually has resolved itself and anxiety has diminished, with a new state of equilibrium having been achieved. Unfortunately, if the crisis has not been resolved in a mentally healthy fashion, the individual proceeds to function on a less mentally healthy level. Caplan and Grunebaum (1967) state:

> It is believed that the methods of crisis resolution used by the individual—whether healthy or maladaptive—will become henceforward a part of his coping repertoire and may be used in dealing with future problems. Thus the individual may emerge from the crisis with increased adaptive capacities and confidence in his ability to tolerate stress and to cope or on the other hand he may emerge with lower adaptive capacities and a greater vulnerability to mental disorder. Therefore, we can say that crises represent mental health turning points. (p. 10)

In delineating a methodology for providing crisis intervention, Caplan formulated the notion of *anticipatory guidance,* the term he used for the preparation of an individual for an impending crisis. This is not always possible because a crisis can arrive unannounced, such as an earthquake, fire, or automobile accident. When the crisis can be anticipated, however, it has been found helpful to provide an emotional inoculation to the subject.

A couple of research studies appear to support this notion. At Yale University, Janis (1958) reported a study of patients awaiting surgery. He found that it was possible to predict ahead of time those who would have the best postoperative adjustment. It appeared that patients who were moderately worried about the surgery, and therefore asked a number of questions, did better in handling the stress than those who appeared to deny any worry and feigned unconcern. From this study comes the original idea of the *emotional inoculation,* wherein individuals are told in some detail of the emotional and physical experiences they can expect. A similar and dramatic study conducted by Egbert, Battit, Welch, and Bartlett (1964) at Massachusetts General Hospital found that patients who had been prepared for surgery through detailed information and instructions for coping with the postoperative experience not only required less postoperative narcotic medication but were ready for discharge on an average of 2.7 days before the contrast group of patients who did not receive this kind of preparation.

It is not a difficult conceptual transition from these studies to the rationale behind the present research. We sought to provide crisis intervention, including anticipatory guidance, to patients who were undergoing coronary bypass surgery. The counseling provided for our surgical patients was based on Caplan's maxims for the provision of crisis intervention.

For example, he believed that the maximum benefit can be gained from crisis intervention if it is delivered at short intervals rather than extending over a period of several months; and in our study, we followed coronary bypass patients for 8 weeks after discharge.

Crisis intervention theory suggests that the integrity of the family should be supported by providing assistance to the family in their own home whenever possible. The crisis intervention counselor visited patients in their homes after surgery and attempted to help the family provide as much support for the patient as possible.

Also, utilization of the already existing social support system is a maxim of crisis intervention; and our counselor was able to encourage the active support of clergy, family, nursing staff, and organized self-help support groups such as the Zipper Club. Crisis intervention counseling hopes to maximize its effectiveness through capitalizing on the already existing network of helpers available to an individual in his day-to-day life, as these are part of his normal coping strategies.

Caplan noted that one of the characteristics of the patient–counselor relationship in crisis intervention is an immediate bonding, which results in patients' becoming very dependent upon the provider of crisis intervention. This was noticeable with our patients primarily during the hospital period, especially the few days immediately following surgery. Crisis intervention theory, however, postulates that this dependency does not last, is of short duration, and is followed by resumption of independence on the part of the patient. We found this to be true.

One of the goals of crisis intervention is educational and informational in that it seeks to enable an individual to master a problem conceptually. Providing information regarding their surgical experience and hospital routine to the patients and their families was an important part of the counselor's role and one which seemed to be uniquely suitable for someone who was not directly part of the surgical team or medical staff.

Perhaps one of the most important aspects of crisis intervention counseling is the one which is the most difficult to follow. This is the emphasis placed on dealing with the present mental health problem rather than dealing with past emotional problems carried into the present crisis. Although an individual may appear to bring to the crisis maladaptive coping styles that stem from unsuccessful past experiences, it has been found that it is possible to encourage alternative, more productive coping styles during a crisis. This can enable the individual to move away from his earlier maladaptive efforts and incorporate more successful coping strategies into his repertoire for the future. Our patients were encouraged to deal with the "here and now" problems associated with recovery from coronary bypass surgery rather than grappling with unresolved emotional baggage.

Bruce Baldwin of the University of North Carolina at Chapel Hill Department of Psychiatry has done much to systematize the delivery of crisis intervention services through seminars, papers, and teaching graduate students. He stresses that crisis intervention as a model is neither a theory of personality nor a comprehensive theory of psychotherapy but that it is a limited but "important framework for responding to a normative life event: the emotional crisis."

Baldwin (1980) lists a number of myths surrounding crisis intervention which he would like to dispel. He feels that it is too often considered a technique only suitable for responding to emergencies; that it is thought of as a "one-shot" form of therapy; or that it is only a holding action until long-term therapy can begin. He also feels that it is not believed that crisis intervention can produce lasting changes, and he is particularly concerned about the myth that no special skills are required to be a successful and effective crisis intervention counselor. Regarding the last myth, Baldwin stresses that a crisis intervention counselor must have good conceptual skills in order to understand the patient's problem quickly and to begin to develop strategies for change, and that the crisis intervention counselor must have the clinical skills necessary for implementing an effective therapeutic strategy as well as good communication skills in order to exchange information accurately with the client.

Crisis intervention postulates that *homeostasis* is a state of psychological comfort which is the very subjectively perceived normal state of a person. Throughout development, an individual is learning coping mechanisms that help maintain this comfortable level of balance. During an emotional crisis there is a breakdown of effective coping responses for a number of possible reasons (Baldwin, 1980).

1. It is possible that the individual does not have a sufficient repertoire of coping behaviors to respond to his present stressful situation. For example, most of our male patients lacked appropriate coping strategies for dealing with a surgical procedure and the coping strategies they attempted were often maladaptive.

2. The individual may have learned maladaptive responses previously and is using them again inappropriately in this crisis. This would be the case if a patient had had a previous experience with medical personnel which had turned out badly. As a result, he assumed a combative posture with the coronary bypass team immediately on the basis of his previous negative experience.

3. Possibly the individual has experienced more than one highly stressful event in close temporal proximity, and that has overwhelmed his coping capabilities. He might have been able to cope with each stressful situation individually but could not handle them in combination. For many of our patients, heart attacks precipitated surgery. The heart attack had followed a number of stressful life events and therefore the coronary bypass surgery itself was a culmination of stressful experiences.

4. The stressful event may reactivate a previously unresolved conflict for which no adequate coping behavior is available. Baldwin calls this a *precipitant*—a thought or feeling that is aroused during a crisis that usually has some past experiential component. Some patients had lost their fathers to heart attacks when their fathers were the same age; and the feelings associated with this experience continued to interfere with their capability to be objective about their own condition.

5. The individual may already be under stress which is causing him

to stretch his coping capacities to the limit; therefore an apparently small, insignificant stress can overwhelm him. Most patients had long histories of angina which had required them to restrict their activities and which had provided them with a constant reminder of their own mortality. Even though the bypass surgery was undertaken with the expectation of alleviating the angina, some patients were too "worn down" from their long experience with pain to be capable of handling the added stress of the surgical procedure.

6. The crisis may represent a situation never encountered by the individual before. This might be a first experience with hospitalization for some of the patients whose coronary bypass surgery was necessitated by a severe heart attack.

7. Coping behaviors which are suitable for one social context may turn out to be inappropriate for another social situation. Men who had been successful executives, accustomed to issuing orders, found it very difficult to assume the passive role of the "good" hospital patient.

Baldwin provides some interesting speculation about why individuals utilize maladaptive coping behaviors which are not effective in ameliorating the emotional pain of the crisis situation. He suggests that possibly the old ways of coping are familiar or safer; possibly there are secondary gains in maladaptive behavior; an individual may be too emotionally overwhelmed to seek alternative coping strategies or may not have had exposure to alternative coping styles. If an individual appears to be avoiding the utilization of a more adaptive coping style, it is important to find out why the person is afraid to take the risk of a new pattern of behavior. In order to do this the crisis counselor could utilize a structural model which lists the things obtained by the individual when he persists in his maladaptive response as well as the things avoided by its utilization.

If the postcrisis resolution is detrimental, or, in other words, if the person has resumed a level of functioning which is less adaptive than it had been prior to the crisis, it is possible that a precipitant is responsible. This precipitant causes behavior which is not constructive for adequate crisis resolution. For Baldwin, a precipitant is a negative coping response and may include features such as having felt similarly at an earlier time when in emotional distress, or a feeling that the scenario has occurred before, or that there is some similarity between the personalities involved and old behavior patterns. In the present study, for example, the authoritarian position of medical doctors can elicit old patterns of coping with authority. On the other hand, the crisis event may become a crisis because it occurred on the anniversary of another unresolved event in the patient's life. An example would be surgery occurring at the same age at which a parent had died or suffered a heart attack.

The following section will provide a detailed description of the utilization of crisis intervention with coronary bypass surgery patients. The section should be considered a general guideline for anyone interested in providing this type of intervention to similar patients. It is not a description of the method used in the research study itself, since those details will be provided in the general method section. The next section derives from the clinical experience of the crisis intervention counselor and includes a summary of her general conclusions and recommendations after two years of work with coronary bypass surgery patients.

CRISIS INTERVENTION APPLIED IN CORONARY BYPASS CASES

Applying the crisis intervention modality to the open-heart surgery patient appears to be a natural and realistic approach. Working with a patient for approximately 10 weeks encompasses the presurgery stage, the postoperative stage in hospital, and the initial 2 months of recuperation at home. The manner in which a patient copes both prior to surgery and after surgery will obviously influence his recuperation and perhaps even his very survival. There is no way to eliminate the stress and anxiety caused by coronary bypass surgery, but it is possible to help the patient cope with both in a constructive rather than a destructive manner.

In the case of the coronary bypass patient, it is important to understand that there is no one appropriate coping style. Most patients will use several coping mechanisms during different stages of recuperation. The counselor must be able to assess the coping patterns accurately and to determine the effect these behavior patterns are having on both the patient and the family. Specific coping styles will be addressed further on in this chapter.

The physical recuperation period of the open-heart surgery patient varies from individual to individual. The emotional recuperative period will also vary. Therefore it is imperative to reassess the typical crisis intervention model and to provide for a longer period of counseling after working with the patient and family weekly for the initial 8 weeks after discharge. Meetings spaced at 2-week intervals are advisable until the patient and family experience a return to their normal pattern of psychological functioning. This assures that the counselor has been able to assess diagnostically the emotional health of both the patient and family system prior to the medical crisis. If there were psychological problems and marital and family tensions evident before the crisis, then crisis intervention work by strict definition would not include remediation. Prior tensions are usually exacerbated by the open-heart surgery experience. Working with this intensification is the goal of the crisis intervention counselor. Remediation of the prior existing problems gets into the area of long-term therapy and is beyond the scope of crisis intervention. This is not to say that there is not usually a serendipitous effect upon a patient and the family. If the patient and the family learn how to cope constructively with a crisis, this achievement should certainly influence long-term personal interactions.

The support system of the bypass patient usually centers on a key family member or a close friend. An adequate support system is essential for a patient's psychological adjustment. The counselor must be able not only to assess the strength of the support system but also to work with this person in helping the patient during his recuperation. As with coping mechanisms, support systems can be either facilitative or destructive. The most delicate task is to help the family member(s) or friend to understand the patient's behavior and to accept the sometimes drastic mood swings and temporary personality changes. The family member experiences intense anxiety and may often exacerbate a patient's stress.

It is important to recognize that the family member usually suffers from tension almost equal to the patient's tension. The difference is usually not in the degree of stress, but in the manner in which it is exhibited. Adding to the family member's stress is the fact that it is the patient who is receiving support and nurturance from family, friends, and medical personnel. The family member is asked to provide a bulwark for the patient while often feeling a loss of emotional support from the patient, who must conserve both physical and emotional energy. A full discussion of the psychological ramifications upon the family system will come later.

MEDICAL STAGES OF CORONARY BYPASS SURGERY

Preoperative Stage

A person who is scheduled for open-heart surgery has first undergone a cardiac catheterization. This procedure may have taken place several days prior to surgery, or the patient may have been discharged from the hospital after the catheterization and then readmitted at a later time for the surgery. Prior to the catheterization, there may have been years of anginal pain and resultant fears suffered by the patient. In other cases, the knowledge of a heart condition may have come upon the patient suddenly and without warning. Normally, a patient enters the hospital 2 days before surgery. A multitude of tests are run during this preoperative period: blood studies, chemical profiles, urinalysis, chest x-ray, electrocardiogram, and sometimes pulmonary function tests.

The patient is taught certain breathing exercises and is instructed in the necessity of coughing after surgery. Sometimes breathing treatments are ordered. The surgeon, cardiologist, and the anesthesiologist meet with the patient prior to surgery. In some hospitals, a nurse will teach the patient about the surgery and answer further questions. The afternoon before the operation, the patient is prepared for the procedure. This preparation consists of a shaving of the body. The patient is asked to send his belongings home, since after surgery he will be moved to the surgical intensive care unit.

After Surgery

A patient spends approximately 2 hours in the recovery room. During this time the endotracheal tube may be removed. During the time in the recovery room the patient will be carefully monitored and an assessment of his reactivity observed.

After the recovery room, the patient is moved to the surgical intensive care unit. Here he usually remains for 2 or 3 days. The patient is monitored constantly for EKG, blood pressure, temperature, venous pressure, urinalysis, fluid intake and output, and daily weight. The chest tubes and drainage catheter are usually removed after 24 hr. The patient is encouraged to move and to cough and usually has breathing exercises every couple of hours. By the second day in intensive care, the patient is encouraged to sit in a chair for short periods. Fluids are usually given the morning after surgery, and foods by the end of the day. The intensive care period can be a disturbing time for both the patient and family, as the constant activity of lights, alarms, computers, and noises of the oxygen and suction catheters can be unnerving.

When the intensive care phase is over, the patient is then moved to a floor where he is monitored by telemetry. This constant monitoring of the heart takes place for approximately 4 days. During the telemetry phase, the patient first begins to walk around his room and then into the hallways. Intravenous and oxygen apparatus are usually removed during this stage. A low-salt diet and fluid restrictions are in effect at this time.

After the telemetry monitor is removed, the patient is in his last phase of the hospital recuperative period. He is allowed to take a shower on his own and is encouraged to walk in the hallway several times a day. At this stage, the patient may not have his appetite back and may still have difficulty with pain. The patient is usually discharged 10–14 days after surgery.

The usual regimen for doctor's appointments is to see the cardiologist about a week after the arrival at home and then to have appointments spaced at the cardiologist's discretion. Appointments with the surgeon are generally 3 weeks after surgery and then 3 months, 6 months, and 1 year later. During the first few weeks at home, the patient is encouraged to walk specified distances several times a day and to get plenty of rest as well. The person who is planning to return to work can usually do so part time at approximately 2 months. This return to work course is variable and depends on the person's state of health as well as his mode of employment.

PSYCHOLOGICAL STAGES OF HEART SURGERY PATIENTS

In this section, the various possibilities of psychological sequelae will be discussed, and in later sections the specifics of dealing with these situations will be outlined. For the purpose of clarification, the psychological stages will be delineated in terms of the medical time phases. It should be understood that these psychological attitudes may overlap or repeat themselves during several phases.

Before Surgery

Anxiety is the universal emotion prior to undergoing open-heart surgery. Some patients manifest anxiety outwardly, whereas others have the emotion so well under control that it is difficult to perceive. Anxiety before surgery is normal; it is rather the degree and manner in which it is manifested that is important. If the anxiety level of the patient is incapacitating and having an effect on the physical condition, then counseling may help to ameliorate the situation and lessen the stress.

Depression is sometimes noted in a patient before surgery, although it is more commonly found in the postoperative phase. The depressed patient will usually have difficulty sleeping, eating (either loss of appetite or excessive appetite), and concentrating. The depressed patient may also have difficulty with interpersonal relationships; he may be withdrawn and uncommunicative. This is the most difficult psychological manifestation to deal with and also the most serious. A depressed patient may be signaling feelings of futility and a desire to give up. Depression prior to surgery may drain the person of the will to survive which is so vital to the surgical recuperative outcome.

Anger is often expressed or felt by patients. Although the external manifestation might be anger, often the internal emotion is fear. It is usually more acceptable culturally for a male to express anger, whereas fear is not so easy for some males to express. Criticism of medical personnel, the hospital, or the state of the world in general often masks an underlying fear of the surgical outcome. Sometimes a patient will even direct this anger against himself and blame himself for his physical situation beyond what is realistically within his control. Often the patient will project this anger onto a spouse, and a significant erosion of the marital relationship takes place. Most patients need to find a reason for the medical state in which they find themselves. Even if the rationale is not scientifically valid, a patient will often cling to the reason in order to provide some structure for himself.

The characteristics of the Type A personality are often noted in a person before surgery. This personality pattern is best described as an aggressive, incessant struggle to achieve as much as possible in as little time as possible. This type of behavior is most socially acceptable and is given status in our society. Working with a patient who displays this pattern requires extreme patience, for it is not easily changed. Also, it can create problems for the patient in a hospital setting where passive, cooperative behavior is rewarded.

Bypass surgery often provides the setting for latent fears to come to the surface. Death and all its physical, psychological, and religious connotations are difficult to deny in the preoperative stage. The very knowledge that one's heart is stopped during the surgery can be an awesome and frightening concept for a patient. Not every patient verbalizes these fears directly, but almost everyone struggles internally with them. Seriously disturbed patients who are suffering with depression may also harbor a death wish. The counselor must be astute enough to pick up on subtle clues coming from the patient's verbal or nonverbal communications.

Denial is also often seen before surgery. A majority of people are conditioned to be strong and controlled in the face of a crisis. There is hypothesized a healthy degree of denial which does enable a person to cope with stress, a certain protective device which everyone needs when facing a crisis. However, if denial of a person's medical condition persists into the recuperative period, then it has the potential of interfering with the rehabilitation process and with the possibly necessary alteration of one's lifestyle. Excessive denial before surgery often results in an angry patient afterward.

In the Surgical Intensive Care Unit

A rare but frightening phenomenon occurring during this stage is a type of psychosis brought on by sensory deprivation, sleep deprivation, certain medications, and stress reaction following the surgery. This form of disorientation usually dissipates once the patient leaves the intensive care unit. Sometimes the syndrome lingers and it is not until the patient is in his home surroundings that he is once again totally in touch with reality. Along with this psychosis may also appear a paranoid reaction. The patient may believe that the medical personnel are conspiring to harm him. Some patients become disoriented in time, place, or person. If a patient is experiencing these symptoms, he needs constant verbal and tactile reassurance. If he is aware enough to know that he is not himself, he must be encouraged to talk about his fears and helped to realize that his psychological state is a temporary one.

Anger may also be exhibited during this stage, because the patient feels totally at the mercy of machines and medical personnel. He may tend to feel a total lack of control over his person, and sometimes the only control he feels he can harness is his anger.

At the opposite end of the spectrum is the patient who becomes extremely passive and withdrawn. He has given up the control, and although he may be an easier patient to handle, his behavior is more alarming than is that of the angry patient.

After Intensive Care

Most patients react positively to the move out of intensive care. A few become more anxious as a result of knowing that they are not being monitored so thoroughly. It is as if their security had been taken from them. Some patients begin to experience fear as they begin to walk on their own and regain some of their independence. There now begins the questioning in the patient's mind of how successful the surgery was and the anticipation of a resumption of his former life-style. Often at this stage patients will become very introspective and begin to think about reordering their priorities. Life suddenly becomes more precious, and patients may exhibit periods of nostalgia and sentimentality. Often patients will comment that they seem to cry very easily, and when this has not been a familiar behavior pattern it can be frightening for both patient and family.

Depression is the most common psychological state during this phase. The discomfort and weakness that the patient experiences often leave him for a time feeling worse than he did before surgery. Thus it is difficult at this stage for the patient truly to believe that he is doing well and that the surgery was in fact successful. The overall physical and mental stress that the patient has endured also plays a significant part in depression. This depression can be thought of as the way in which both body and mind are responding to stress overload. There is a significant time variance in the longevity of depression, but if this state endures for longer than several weeks, it is important for the psychological counseling to continue. As the period of depression comes to an end, there is usually an ebb-and-flow factor: days when the person feels more cheerful interspersed with days of depression.

After Discharge

The initial week at home is often more difficult for the patient than his stay in the hospital. The patient may feel weaker and more fearful without the support system in the hospital. Counseling the patient on realistic expectations of the recuperative period helps to alleviate the anxiety of leaving the hospital. If the patient has a supportive family that understands his emotional needs, the transition from hospital to home is easier. If there has been prior stress in a marital or family relationship, then often this stress is exacerbated upon the patient's homecoming. If the patient is a man, then his dependency and the resulting role reversal may create friction. This same situation may occur with a woman who is accustomed to independence. Often as the patient begins to grow stronger and assume his or her previous role in the family, the spouse may feel less needed and have difficulty in readjusting to the prior role. Couples who have healthy relationships prior to surgery usually grow closer as a result of the medical crisis.

Realistic expectations are one of the most important factors insuring a healthy emotional state. If the patient can allow himself the time it takes to regain his strength and to test the success of the surgery, then his psychological well-being is greatly insured. Conversely, if the patient is impatient with himself and unrealistic in his goal, then the stage is set for depression. Patients who have jobs to return to are usually more strongly motivated toward recovery. If the person is retired it is important for him to have a meaningful goal or purpose.

During the first few weeks at home, the patient will have periods of emotional highs and lows. This is a normal phenomenon. He may feel like withdrawing at times and may not be receptive to socializing in the beginning. It is almost as if he needed to draw a protective shield around himself as he regains his strength and self-confidence. Patients' families often comment at this point that there has been personality change. The person may become very sensitive as well as extremely critical of family members. There may be less tolerance for others' shortcomings as well as his own. The tolerance level is often at the same low level as is the energy level. Reestablishing a couple's sexual relationship also takes time. The patient may be unsure of his physical stamina and afraid to have intercourse. The spouse may also be fearful for the patient's well-being. This protective syndrome may continue far beyond the normal limits and become a source of dissension between the couple.

As the weeks of recuperation progress, the discomfort diminishes, strength is restored, and the patient will begin to feel his depression and anxiety lifting. The best sign of emotional health is when the patient begins to make plans to resume his work or, if retired, to resume a favorite pastime. As the individual begins to assume more control over his life, his spirits revive.

There are some patients who have a tendency to become "cardiac cripples." These people become so bound by fear that they regress into a state of near-helplessness. This helplessness is based upon a serious psychological condition and must be treated with a well-planned counseling approach.

COUNSELING AND ASSESSMENT

Before Surgery

Counseling intervention is most helpful if initiated as soon as the patient is identified as a surgical candidate. The therapeutic relationship between the counselor and patient is developed more naturally and the counseling intervention is utilized more efficiently at an early stage. Several short sessions with the patient before surgery are more useful than one long, intense session the day before. It is helpful for the counselor to meet with the family as well. The family, or a close friend, can often provide information which will help in additional diagnostic insight. The patient often expresses a desire for the counselor to meet with his family as a way of providing additional support. Many patients are very concerned about the stress on their families. Thus the counselor's willingness to include the family in the treatment plan is usually reassuring to the patient. Utilizing several meetings before surgery provides the counselor with the opportunity to see the patient alone, see the family alone, and then see them together.

The intent of the counselor should be to provide emotional support for the patient. This entails both imparting relevant information and allowing the patient a secure setting in which to discuss his feelings. Often the patient is protective of his family and will not allow himself to disclose fears or anxieties to them. The counselor must quickly provide the patient with permission to ventilate his feelings and concerns. This very act of catharsis may often be enough to alter significantly a depressive or anxious state. The counselor must be sufficiently well trained and intuitive to assess the ego strength of the patient. It is just as important to know the emotional areas that should be avoided. For example, a well-defended individual practicing denial often needs to retain this state of denial before surgery. Tampering with the patient's defenses can be destructive of his emotional and medical well-being. Supporting these defenses is often an integral part of presurgical counseling. The counselor must be sufficiently skilled to make this determination quickly and accurately.

The most important areas to discuss with a patient and his family before surgery are as follows:

1. Surgical intensive care unit and its emotional effects
2. Possible temporary disorientation after surgery
3. Depression in the recuperative phase

The most important areas to assess in a patient before surgery for the purpose of diagnosis and treatment are as follows:

1. Past medical history
2. Past psychological history
3. Major stress areas in the patient's life and previous coping behaviors
4. Present coping behavior pattern
5. Patient's support system (e.g., family) and the state of the relationships
6. Anxiety level
7. Depression level
8. Degree of positive thinking and will to survive
9. Latent death wishes or suicidal ideation
10. Assessment of patient's degree of need to control his environment

The above are clinical judgments to be made on the basis of interview assessment techniques and psychological testing.

In the Surgical Intensive Care Unit

Supportive counseling is essential during the intensive care stage. As we discussed above, the patient is exposed to a host of frightening sequelae during this stage. The counselor's main intent must be to help the patient adapt successfully to this critical illness phase. The following are prerequisites for successful adaptation by the patient:

1. Ability to allow control to be taken from him for a time
2. Ability to maintain supportive defenses against the stresses of the unit
3. Ability to communicate his needs and his feelings (verbally or non-verbally)
4. Trust in the medical personnel

The counselor must be facilitative in these four areas.

Since time spent with the patient is by necessity minimal during this stage, the counselor often will act as an intermediary between the patient and medical personnel, helping to sensitize the latter to the patient's special emotional needs. Short daily visits with the patient are advisable. The counselor must always keep in mind that the patient may be cognizant of communication around him even when he appears nonresponsive. If a patient is disoriented or evidences "ICU psychosis," the counselor can be helpful in reassuring both the patient and family during this transient, albeit frightening, time.

After Intensive Care

The transition from the security of the intensive care unit (ICU) to the more independent stage of telemetry in another area of the hospital may be disconcerting to the patient. Because the patient is expected to do more on his own, both fear and frustration can develop. The counselor must be alert to emotional nuances during this stage. This is often the time when depression, anger, or intense anxiety can arise. It is often difficult to distinguish between the obvious physical fatigue and weakness a patient feels and depression. The best indices are the patient's affect, his verbalizations, and his willingness to try to become more mobile. A recalcitrant patient who is angrily refusing medication or treatment is often masking underlying anxiety. A certain degree of depression and anxiety are normal. The counselor must be able to determine when these factors are interfering with the recuperative process. Patients will respond differently, and it is at this point that the presurgical psychological assessment becomes a valuable tool. If the counselor has established a good relationship with the patient, the patient will usually be very willing to talk about his feelings at this point. Often patients must go through the catharsis of relating their ICU experiences. Conversely, some patients need to be given permission to forget this stage. A few patients will have no recollection of their days in ICU and will need help in reconstructing this period for the sake of a better feeling of control.

It is only after working with several patients that the counselor will begin to feel more comfortable with the medical and psychological stages after surgery. The counselor must not act merely as a friendly visitor during this stage, but rather inspire confidence in the patient that the counselor can serve to help him work through any emotional problems. It is too simplistic to slip into an ineffectual role. The counselor has a responsibility to observe the patient's coping behavior and to facilitate progress through the emotional stages.

PREPARATION FOR DISCHARGE AND POSTDISCHARGE COUNSELING

An ideal program for counseling the open-heart surgery patient includes emotional support after hospital discharge. This counseling is offered as part of a hospital's outpatient services. The patient arranges an appointment with the counselor for approximately 2 weeks after discharge, during which the counselor and the patient evaluate the patient's emotional state. Points to be included in the evaluation are as follows:

1. Evidence of depression and/or anxiety
2. Adjustment to the home situation
3. State of marital or other family relationships
4. Adherence to physician's orders and rehabilitation protocol
5. Ability to cope with diminished physical abilities and postsurgery discomfort
6. Assessment of future-oriented goals (e.g., return to work, extracurricular activities, travel)

The patient should be encouraged to telephone the counselor sometime during the first week at home. Leaving the hospital often produces an insecure feeling in the patient as the hospital support system is left behind. The counselor can act as a buffer in this transition. Often the patient and the family will have formed a solid bond with the counselor, and the "letting-go" process must be gradual and based on a clinical assessment of emotional needs. The dependency issue must be addressed several weeks after discharge, and the patient must be made aware that the counseling relationship is being supplanted by an increasing self-confidence on the patient's part. The counselor must understand the deleterious effect of allowing dependency to continue beyond the necessary time. The welfare of the patient must constantly be assessed. Several postdischarge visits can be planned with the patient over a period of 6–8 weeks. If the patient is to return to work, there may be much anxiety related to this issue which the counselor can effectively handle with the patient. Having been placed in the category of cardiac patient often produces a loss of self-esteem which directly manifests itself in the return-to-work period. Marital disruptions are also common, and the counselor must be alert to this possibility. The patient who is in the dependent role is often uncomfortable, and if the patient is a man, this role reversal can be most threatening. If marital problems arise, then conjoint counseling is essential.

When the counseling relationship is terminated, the patient must be made to feel free to telephone the counselor if necessary. The control should be left with the patient, for this gesture is a reaffirmation of the patient's return to self-sufficiency.

PREREQUISITES, TRAINING, AND SUPERVISION OF THE COUNSELOR

The person selected to counsel open-heart surgery patients must be well grounded in human relations and psychodynamics. A graduate degree of at least the master's level should be a requirement. The area of speciality may be either psychology, psychiatry, psychiatric nursing, or social work. This person should also have a minimum of two years' experience in counseling. Prior experience with medical patients is helpful but not essential. The skills required for counseling open-heart surgery patients are as follows:

1. An understanding from the medical and psychological perspective of the stages of open-heart surgery
2. Knowledge of medical terminology and hospital routine
3. Ability to work individually with patients and experience in marital and family counseling
4. Experience with the crisis intervention approach to counseling

5. Ability to assess patients clinically using both interviews and psychological testing
6. Ability to communicate effectively with medical personnel
7. Knowledge of medical charting and ability to write psychosocial reports
8. Emotional maturity to be able to handle the crisis environment of a hospital and its personal effect on the counselor
9. A degree of comfort concerning the topic of death
10. Ability to be both empathic and objective when working with patients

These are the basic skills required for such counseling. Adequate training and good supervision will serve to refine these skills further. Before a counselor begins to work with patients, he should familiarize himself with the hospital environment. The counselor should learn the physical facilities of the hospital and talk with as many physicians, nurses, and hospital personnel as possible. Informal interviews with open-heart surgery patients should be made available to the counselor. These interviews should encompass all stages of the procedure. Two weeks invested in this manner will serve to provide a better foundation and a greater sensitivity for the counselor.

There must be weekly supervisory meetings set up for the counselor. The supervisor should be a mental health professional who has had prior experience with open-heart surgery patients. The counselor should be encouraged to present actual case histories on a weekly basis. Tape-recorded sessions with patients will provide an excellent basis for supervision. The counselor must be encouraged to utilize the supervisor's help between appointed meetings as it becomes necessary. The counselor must also have access to a physician who will serve as an advisor to help the counselor understand more fully the medical aspects of open-heart surgery.

REFERENCES

Baldwin, B. *Crisis intervention workshop*. Presented at Chapel Hill, N.C., April 24–25, 1980.

Caplan, G. *Principles of Preventive Psychiatry*. New York: Basic Books, 1964.

Caplan, G., & Grunebaum, H. Perspectives on primary prevention: A review. *Archives of General Psychiatry*, 1967, *17*.

Egbert, L. D., Battit, G. E., Welch, C. E., & Bartlett, M. K. Reduction of postoperative pain by encouragement and instruction of patients. *New England Journal of Medicine*, 1964, *270*, 825–827.

Janis, I. L. *Psychological stress: Psychoanalytic and behavioral studies of surgical patients.* New York: Wiley, 1958.

Lindemann, E. Symptomatology and management of acute grief. *American Journal of Psychiatry*, 1944, *101*, 141–148.

III

Research on Crisis Intervention and Coronary Bypass Surgery

6

Crisis Intervention and Coronary Bypass Patients

Patient Characteristics, Methodology, and Research Design

JOSEPH R. FEIST

The research project that forms the foundation of this volume is unique. Very few surgical studies of the psychological well-being of patients have included any out-of-hospital follow-up. The present study included a significant follow-up for a large number of patients 12 weeks after discharge and for a smaller number of patients 3 years after discharge.

Another aspect of the current study which distinguishes it from other psychological surgery studies is the inclusion of a substantial psychological treatment component. Few studies have provided patients with any treatment, and those that have provided intervention often limited it to relaxation techniques or presentation of information. The present study includes crisis intervention counseling, an approach which seems especially well suited to the emotional and logistical demands of patients undergoing major surgery.

This chapter describes the methodology used in the research project and is followed by a chapter in which the major research findings are presented. It describes the patients who were included in the study, outlines the basic procedure employed, and indicates which psychological tests and other evaluation methods were used to assess the results. The major purpose of the study was to explore the helpfulness of crisis intervention counseling in alleviating postsurgical depression in patients undergoing coronary bypass surgery. The study was conducted over a 3-

JOSEPH R. FEIST • Pimm Consultants, 2699 S. Bayshore Drive, Miami, Florida 33133.

year period and included patients who had surgery at one of two cooperating hospitals: the Miami Heart Institute, Miami Beach, Florida and South Miami Hospital, South Miami, Florida. The research was supported during the first year by a grant-in-aid from the American Heart Association, Miami Chapter. The subsequent years of the project were funded through the generosity of the Research Department of the Miami Heart Institute.

THE PATIENTS

In order to allow the research team to follow patients undergoing coronary bypass surgery before surgery and for a substantial number of months following their discharge from the hospital, it was necessary to devise a suitable method for selecting appropriate research subjects with this availability. The research team began by identifying the most typical sequence of events which leads to a patient's decision to have coronary bypass surgery. It was found that most commonly the surgery was elective rather than emergency in nature. The patient had usually conferred with a cardiologist, had undergone angiography at the catheterization laboratory, had consulted a thoracic surgeon, and then had made appropriate arrangements with a hospital. The team chose the surgeon's office as the best "patient window," the place from which it could identify patients who might be recruited for this study. Although it might have been desirable to pick up patients following angiography, the surgeon's office was selected largely for pragmatic purposes. By obtaining from the surgeon's office the names of candidates already scheduled for bypass surgery in the forthcoming week to 10 days, arrangements could be made to recruit patients without losing too many. If we had chosen instead the cardiac catheterization laboratory, individuals learning of their disease but eventually choosing nonsurgical treatment might have been approached and then found ineligible for inclusion in the study. It would have been interesting to examine patients psychologically at this point in order to assess whether or not those opting for surgery are different from those who select nonsurgical interventions and to assess the impact that the lab results had on their mental status. The research team decided, however, that it was more feasible to recruit subjects closer to the final surgery decision point.

The study was described to a group of surgeons whose practice included large numbers of coronary bypass operations. Following their agreement to participate, arrangements were made with the appropriate office staff to provide the research team with the names of patients scheduled for bypass surgery. Therefore, the initial research subject pool came from unselected, consecutive coronary bypass surgery cases referred from a private group of five thoracic surgeons who operated in two local private hospitals.

Before a patient was actually included in the study, other eligibility requirements were considered. Most of those who undergo coronary bypass surgery are men. We were particularly interested in possible interactions between personality variables and psychological outcome following surgery. We decided to exclude female patients and instead focus our attention on the rather typical case of a male head of household who had to relinquish this role, or at least modify it substantially, during the time of recovery and then reestablish himself afterward. Further, in discussions with medical advisors we were frequently told that female surgical candidates appeared to be quite different in personality from the more usual male surgical cases. In an effort to simplify this preliminary investigation, women were excluded but future research should certainly include them. Other eligibility criteria were necessary because we were interested in following patients for a significant period of time following their hospital discharge. Therefore we limited our study to those patients who were living nearby, hoping that this group would remain accessible to us so that the study could be completed as it was designed. We limited the study to English-speaking surgical candidates. Miami is a bilingual community where Spanish is commonly spoken. Because our counselors were not bilingual and the test instruments were written and normed on English-speaking subjects, we felt that this criterion made sense. Finally, it was necessary to obtain the informed consent of patients before they could be included. Patients were contacted by the research team in the hospital on the day before surgery. Arrangements were made to explain the purposes of the study and outline what a patient's participation might require. Most patients readily agreed to participate in our psychological study, which might involve them in counseling (depending upon later group assignment). Patients frequently commented that they would agree to participate, not for themselves but "to help others." In summary, the patients in our study were volunteer male local residents who spoke English and who were about to undergo coronary bypass surgery. They were selected from a patient pool of consecutive surgical cases scheduled for coronary bypass surgery by a local team of five thoracic surgeons. Inclusion in the study involved neither extra fees charged to the patient nor compensation paid to the patient for his participation. Usual medical, surgical, and nursing procedures were followed.

Altogether, 104 patients were selected in this manner and are in-

Table 1. Demographic Summary

Total number of patients	104
Ages of patients	
Range	32–75 years
Mean (average)	59.3 years
Percentage over 55	72.1%
Percentage 55 or younger	27.9%
Sex of patients	
Males	100%
Females	0%
Patients' marital status	
Married	95.2%
Divorced	2.9%
Widowed	1.0%
Never married	1.0%
Occupational status of patients	
High-level executives, large business managers, professionals	29.2%
Administrative personnel, small businessmen	38.9%
Clerical, sales, or technical workers	15.3%
Skilled or semiskilled workers, machine operators	13.9%
Unskilled workers or dependent	2.8%

cluded in the study. Demographic characteristics are described in Table 1. The presurgical medical characteristics of our research subject patients were obtained from a review of their hospital records. The medical descriptive variables are summarized in Table 2 and are described further below.

MEDICAL DESCRIPTION

Of the patients undergoing coronary bypass surgery in the present study, 51.6% had previously had at least one heart attack and 52.9%, as compared with 50% of the controls, similarly had a positive heart attack history. This difference is not statistically significant. Similarly, there were no significant differences found between treatment and control patients on any of the other presurgical medical variables described below.

Just over one-third of all surgery patients (37.6%) included in the study had a history of hypertension. Slightly more control patients had a

Table 2. Baseline Medical Variables

	All patients	Treatment group	Control group
Number	104	57	47
Age (years)	59.3	59.1	59.6
History of prior infarct	51.6%	52.9%	50.0%
History of hypertension	37.6%	42.9%	33.3%
Family history of heart disease	62.4%	53.2%	73.7%
Angina longer than 24 months	50.0%	49.0%	51.2%
N. Y. Heart Association Functional Class III	50.6%	53.3%	47.4%
Systolic BP greater than 149 mm	14.0%	7.8%	21.4%
Diastolic BP greater than 89 mm	26.9%	19.6%	35.7%
Triple coronary disease	63.4%	56.9%	71.4%
Ejection fraction less than 45%	17.3%	19.6%	14.3%
Left main coronary disease	33.3%	31.4%	35.7%
Four or more bypasses	33.3%	27.4%	40.5%

hypertension history than did treatment subjects; however, the difference, 42.9%, compared with 33.3%, was not statistically significant.

Interestingly, almost two-thirds of all patients (62.4%) had a family history of heart disease. Despite random assignment of patients to groups, the numbers in treatment and control groups with such a history approached significance; 73.7% of control patients have a positive family history of heart disease, compared to only 53.2% of treatment subjects. These groups do not differ on Beck Depression Inventory scores preoperatively, however.

Almost two-thirds of the patients in the study (63.4%) had triple coronary disease. Some 56.9% of treatment patients and 71.4% of controls had this diagnosis. These patients do not differ significantly on Beck Depression Inventory scores preoperatively.

One-third of the patients in the research project (33.3%) had left main coronary disease: 31.4% of the treatment patients and 35.7% of the controls, a difference which is not statistically significant.

Only a very few patients had previously had open-heart surgery (10.8%)—slightly more in the treatment group, 13.7%, than in the control group, 7.1%. This difference is not statistically significant.

With regard to angina history, nearly all patients had a positive history (96.8%): 100% of the control patients and 94.1% of the treatment patients. According to the medical records, these coronary bypass surgery patients had suffered with angina for an average of slightly over 4½ years prior to surgery (55.12 months). The average number of months of presurgical angina in the treatment group was found to be 52.67, as compared to 58.15 months in the control group.

With regard to the number of bypasses actually grafted during surgery, most patients had either three or four. The actual distribution of patients by number of bypasses is as follows: 1 = 8.6%, 2 = 15.1%, 3 = 41.9%, 4 = 20.4%, 5 = 9.7%, and 6 = 3.2%.

Preoperative adjustment can be described according to the New York Heart Association Scale. This 4-point functional scale is an ordinal measure in which the higher number indicates greater impairment; 1.2% of the patients were in category 1, 14.5% in category 2, 50.6% in category 3, and 33.7% in category 4. The chi-square on the difference between the treatment and control groups was found to be not significant.

According to the medical records, the average left-ventricle ejection fraction prior to surgery was 58.48. The mean in the treatment group was found to be 58.85, as compared to 58.00 in the control group.

Medical records regarding blood pressure readings taken before surgery indicated that the average systolic blood pressure for all patients was 129.49. The treatment group average systolic blood pressure was found to be 125.99 and that of the control group, 133.76. This difference approaches statistical significance, $\chi^2 = 3.63$, $DF = 1$, $p < .06$.

Diastolic blood pressure readings taken before surgery indicated slightly higher average readings in the treatment group. Overall, the average for all patients was found to be 87.40. The average reading for treatment patients is 92.00, compared to 81.81 in the control group. This difference is found to be not significant.

RESEARCH PROCEDURE

The overall strategy of the project was to randomly assign patient volunteers to one of two groups, treatment or control. Treatment was crisis intervention and consisted of one session of presurgical anticipatory guidance. This involved a meeting with the counselor in the hospital on the day prior to surgery for psychological support and then 8 weeks of postdischarge supportive counseling. These latter sessions were typically held in the patient's home. The control group received no special psychological support during their hospital stay nor during their out-of-hospital recuperation period. Both groups received the usual nursing and medical care afforded coronary bypass surgery patients.

In order to evaluate the impact of crisis intervention counseling on depression and other relevant symptoms which might follow surgery, and in order to obtain information about the psychological impact of coronary bypass surgery, a strategy was devised which allowed the research team to assess the patients psychologically at three observation points. The first of these took place on the day before surgery. At that time, nearly all patients were asked to complete a presurgical battery of psychological tests. The second opportunity for assessment occurred several months after the patient's discharge from the hospital. All patients were asked to complete a battery of tests approximately 3 months following their discharge, a time calculated to shortly follow the termination of the crisis intervention counseling for treatment patients and to assess control patients at a comparable interval. Finally, the third observation point was conducted 3 years after hospital discharge. A long-term follow-up group was constituted by selecting a number of patients from a list of those whom we could find, who had rather complete data on the previous test batteries, and who were willing to be assessed once again.

The research team had hoped to measure by experiment whether or not the experience of taking the presurgical psychological test battery in any way affected outcome psychologically. This can be done by including two additional groups, one treatment and one control, which are exactly like other patient groups but which are not asked to complete pretests. Unfortunately, administrative considerations prohibited selecting any of these patients until the second year of the study and also resulted in an exceedingly small number of patients. Preliminary examination of this information suggests that there may indeed be an impact on postsurgical depression if patients are asked to complete a presurgical battery of tests (see Chapter 7). Additional research is needed, however, to examine this possibility with more clarity.

The actual distribution of patients to groups is given in Table 3, and the numbers of patients in the different follow-up groups is shown in Table 4.

As can be seen in Table 3, a significant amount of pretest data was lost. Despite provision of instructions regarding how the test packets would be recovered, many packets were lost, mainly because of the competing demands on patients' time during the hectic day before surgery. Several packets were lost through administrative problems despite several attempts at devising an effective method of picking them up. Logistically, it was not possible to assign someone the task of remaining with the patient until the battery was completed. Examiner-administered instruments were similarly not feasible.

For patients assigned to the treatment group, counseling began on the day before surgery. During the first year of the study, one individual served as the crisis intervention counselor. The counselor was a married,

Table 3. Distribution of Patients to Groups

Total number of patients	104
Assignment to treatment or control groups	
Treatment patients	57
Control patients	47
Number of patients given pretests—results available	
Treatment group	38
Control group	35
Total	73
Number of patients given pretests—results unavailable: not done or lost	
Treatment group	3
Control group	5
Total	8
Number of patients not given pretests	
Treatment group	16
Control group	7
Total	23

female, masters-level counselor in her early thirties. During the second year the project acquired a second female counselor of similar credentials. Although there was some overlap, one counselor typically spent her time at one hospital and the second worked at the other. Both counselors had had several years of experience working with middle-aged and older adults, and both had had previous experience with crisis intervention.

Patients who were randomly assigned to the treatment group met with the crisis intervention counselor after completing the battery of tests and spent one to two hours with her. During this initial session of anticipatory guidance, the counselor tried to provide "emotional in-

Table 4. Patient Follow-up

Number of patients in short-term follow-up group (3 months after discharge)	
Treatment group	55
Control group	47
Total	102
Number of patients in long-term follow-up group (3 years after discharge)	
Treatment group	16
Control group	18
Total	34

oculation" for the patient to the possible psychological stresses and strains which might be associated with the forthcoming crisis of open-heart surgery. Control patients did not have the opportunity to meet with the counselor. Both groups received the usual presurgical information provided by the surgical and nursing staff.

The primary goal of crisis intervention counseling is to assist the person in coping with the psychological components of an immediate crisis and to return him to the level of psychological functioning that existed prior to the crisis period. It is limited in both time and scope and attempts to use the patient's already existing coping mechanisms and his family, his primary social support system. (For a more complete discussion of crisis intervention please refer to Chapter 5.)

The counselor in this study did not routinely interact with patients while they were in the surgical intensive care unit, although there were several instances when the counselor was called upon to be of assistance. A small percentage of patients experienced acute anxiety attacks during this period and there were one or two cases of actual psychotic episodes.

Ordinarily, postsurgical counseling began when the patient was discharged from the intensive care unit and continued on a weekly basis during the patient's hospital stay. Out-of-hospital sessions were usually conducted by the counselor in the patient's home.

EVALUATION METHODS

As outlined above, there were three observation points at which patients were assessed psychologically by the research team. The first of these involved a 2-hour, pencil-and-paper battery of psychological tests given the day before surgery in the hospital. The second assessment point was at 3 months after hospital discharge. At this point, patients were again asked to complete a battery of self-administered psychological tests, most of which were identical to those contained in the pretest packet. At 3 months, an attempt was made to elicit family member's input by means of a Family Questionnaire which the spouse or other family member most intensely involved with the patient was asked to complete. The Family Questionnaire was designed by the research team for this purpose. The family member also completed portions of an already published test (described below). At 3 months the research team also attempted to elicit feedback about the patient's condition from his cardiologist. For logistical reasons this approach had to be abandoned because of the large number of different physicians involved. For the long-term follow-up group, assessment was done by means of a packet of pencil-and-paper tests, again very similar to the test batteries administered previously. In addition, a semistructured interview was conducted by a research assistant who had not previously been involved in the study. Family members were asked again for their assessment of the patient's status, which they provided in written form.

Each test battery was carefully compiled by the research team. The protocols were designed to assess the patient on a number of dimensions which were thought to be important by other researchers and included primarily symptom measures (e.g., depression, our primary target symptom) and measures of various personality dimensions (e.g., locus of control, discussed below). The composition of the components of each test battery is presented in Table 5.

The Beck Depression Inventory (BDI) is a widely used self-report questionnaire that assesses depression (Beck, 1978). It contains 21 item

Table 5. Psychological Measures

Tests administered before surgery

1. Beck Depression Inventory
2. Minnesota Multiphasic Personality Inventory—Depression Scale
3. SCL-90-R
4. Jenkins Activity Survey
5. Conceptual Level Analogy Test
6. Locus of Control Scale
7. Recent Life Changes Questionnaire
8. Millon Behavioral Health Inventory

Tests administered 3 months after discharge

1. Beck Depression Inventory
2. Minnesota Multiphasic Personality Inventory—Depression Scale
3. SCL-90-R
4. Conceptual Level Analogy Test
5. Locus of Control Scale
6. Recent Life Changes Questionnaire (modified instructions)
7. Millon Behavioral Health Inventory
8. Health Fears Questionnaire
9. Adjustments following Heart Surgery Questionnaire
10. Family Member Questionnaire
11. SCL-90-R Analogue (completed by family member)

Tests administered 3 years after discharge

1. Beck Depression Inventory
2. SCL-90-R
3. Jenkins Activity Survey
4. Locus of Control Scale
5. Millon Behavioral Health Inventory
6. SCL-90-R Analogue (completed by family member)

categories, in each of which several statements appear. Within each category, the patient is asked to indicate which of the items best describes his feelings. He selects from the series of statements which range from low to high on the particular aspect of depression assessed by that category (e.g., sadness). Items are weighted from a score of zero (no depression) to three (high depression). Only one score is obtained from each group of statements even if more than one is endorsed, the category score being the highest score obtained per category. The range of scores on the BDI is from 0 to 63. Some of the clusters of items deal with affective and cognitive manifestations of depression, others with somatic or body symptoms. Because many of the items on the BDI are obviously related to depression, a person may well be aware of what the test is designed to measure while he is completing it.

The Minnesota Multiphasic Personality Inventory (MMPI) is a very well-known measure of personality functioning and psychological symptom distress (Dahlstrom, Welsh, & Dahlstrom, 1972). Because of its length, the research team chose to include only those 60 items which comprise the depression scale. As is true of the entire test, these items are all of the true–false variety. Some are obviously related to depression whereas others seem not to be. All questions were determined on an empirical basis to separate and identify depressed patients from those who are not depressed.

The SCL-90-R is a 90-item, self-report symptom checklist, designed to assess common patterns of psychiatric and medical symptoms (Derogatis, 1977). The patient is asked to rate each of 90 symptoms on a five-point scale which ranges from "not at all" to "extremely" (0–4) to indicate how much he has been distressed by the particular symptom within a specified time period. The scale yields scores on nine symptom dimensions: somatization, obsessive-compulsive, interpersonal sensitivity, depression, anxiety, hostility, phobic anxiety, paranoid ideation, and psychoticism. In addition, the SCL-90-R yields three summary scores: a Global Severity Index (GSI), a Positive Symptom Distress Index (PSDI), and a Positive Symptom Total (PST). The test is designed to evaluate the current symptom status of the patient and is not intended to be a measure of personality. Test–retest reliability of the symptom scales is reported in the manual to range from .78 to .90. Norms are available that are appropriate for use with both psychiatric and medical patients, the latter being used in the present study. Although Derogatis (1977) argues that the symptom dimensions are relatively stable, Hoffmann and Overall (1978) contend that the "instrument measures more of a general complaint or general discomfort dimension than distinct dimensions of psychopathology" (p. 1190).

The SCL-90-R Analogue (Derogatis & Mellisaratos, 1976) was included for use as a companion to the SCL-90-R. It consists of several 100mm lines, the zero point of which is labeled "not at all" and the other end "extremely." The test was designed to describe the same psychological symptom dimensions yielded by the SCL-90-R itself. At 3 months, the spouse or other family member most intimately involved with the patient's recovery was asked to evaluate the patient on two symptom dimensions, depression and anxiety. At 3 years, they were asked to assess depression. The person filling out the analogue is given a description of the symptom to be evaluated and simply marks on the line how much of the symptom dimension is displayed by the patient.

The Jenkins Activity Survey (JAS) is the most widely used paper-and-pencil measure of Type A Behavior Pattern (TABP). TABP, which was previously thought of as a personality characteristic, has been shown to be predictive of coronary heart disease (CHD) (Rosenman, Friedman, Straus, Jenkins, Zyzanski, & Wurm, 1964). It is generally accepted that TABP is best evaluated by a structured interview in which the subject is challenged and his response style in reaction to the challenge is analyzed. Since the present study did not permit such an interview, the JAS paper-and-pencil measure was used instead. The test yields an overall Type A or Type B assessment of the patient plus three subscales which have been identified through factor analysis. The subscales are speed and impatience, job involvement, and a hard-driving approach. Chesney and Rosenman (1982) have noted that the Type A scale can identify patients predisposed to CHD, but they note that the relationship is not so strong as that obtained by means of the structured interview. This relative superiority of the interview has been noted by others as well (e.g., Steptoe, 1981). The JAS consists of 52 items through which the patient describes his own behavior.

The Conceptual Level Analogy Test (CLAT) is a 42-item analogy test designed to measure impairment of abstract thinking ability, which is often a sign of brain damage (Willner, 1971). In one study, scores on the CLAT were related to different types of surgical outcome following heart surgery; however, the finding was strongest for valvular patients (Willner, Rabiner, Wisoff, Hartstein, Struve, & Klein, 1976). The absolute number of correct responses is tabulated and then converted into a standard score which can range from 0 to 19. The CLAT was the only test included in the study for which there were right and wrong answers. It probably was the most taxing of all of the tests administered.

The Locus of Control Scale (Rotter, 1960) was developed by Rotter to assess the extent to which an individual feels that he is in control of his life or controlled by external factors. This instrument has been included in a vast number of studies including research investigating recovery from acute myocardial infarction (Cromwell, Butterfield, Brayfield, & Curry, 1977). Patients are presented with a series of paired statements. They are asked to indicate which statement of the pair they believe in more strongly. In the present study the questionnaire was scored such that higher numbers indicate more external locus of control; that is, people with higher scores believed that their lives were most typically controlled by external events.

The Recent Life Changes Questionnaire (RLCQ) is a self-report measure of significant life events which people have experienced over the past 24 months prior to the test's administration (Rahe, 1975). The patient is presented with a list of 55 life events organized into five domains: health, work, home and family, personal and social, and financial. He is asked to indicate which of the events have occurred by checking a box indicating in which 6-month time period (0–6, 7–12, 13–18, or 19–24) the event in question happened. The patient is then asked to estimate on a scale from 1 to 100 how much personal adjustment was required to deal with that particular event. For the purposes of the present study, we examined those events which had occurred within 6 months prior to surgery during the first administration of the scale and those which had occurred since surgery on the second administration of the scale. The RLCQ yields three different scores: first, the number of life events that have occurred within the pertinent time period; second, the total of adjustment scores provided by the patient for the relevent events; and finally, a series of scaled scores for 43 of the items contained on the questionnaire. The sum of the scaled values of those items endorsed by the patient provides an assessment of how people "in general" might be affected by these same life events.

The Millon Behavioral Health Inventory (Millon, Green, & Meagher, 1982) is a 150-item, self-report, true–false questionnaire which was normed and validated on medical patients. It is designed to provide information useful to health care personnel in the diagnosis, treatment, and management of their patients. The test is composed of 20 scales and is organized around the theory of personality developed by Millon (1969). The 20 scales are divided into four categories: Basic Coping Styles, Psychogenic Attitudes, Psychosomatic Correlates Scales, and Prognostic Indices Scales. For a more complete description of the MBHI, the reader is referred to Chapter 4.

The Health Fears and Adjustments following Heart Surgery Questionnaires were devised by Babette-Ann Stanton of Boston University School of Medicine for her work with open-heart surgery patients. The patient is asked to evaluate various fears and adjustments on a five-point ordinal scale.

The Family Member Questionnaire was devised by the research

team for use in this study. It was designed to allow the spouse of the patient to comment on the experience and on the patient's status.

The research project generated a considerable amount of information about the psychological experience of coronary bypass surgery and the helpfulness of crisis intervention. These results are presented in Chapter 7.

REFERENCES

Beck, A. T. *The Beck Depression Inventory*. Philadelphia: Center for Cognitive Therapy, 1978.

Chesney, M. A., & Rosenman, R. H. Type A behavior: Observations on the past decade. *Heart and Lung*, 1982, *11* (1), 12–19.

Cromwell, R. L., Butterfield, E. C., Brayfield, F. M., & Curry, J. J. *Acute myocardial infarction—Reaction and recovery*. St. Louis: Mosby, 1977.

Dahlstrom, W. G., Welsh, G. S., & Dahlstrom, L. *An MMPI handbook: Vol. 7. Clinical interpretation*. Minneapolis: University of Minnesota Press, 1972.

Derogatis, L. R. *SCL-90-R (revised) version manual-I*. Baltimore: Johns Hopkins University School of Medicine, 1977.

Derogatis, L. R., & Millisarotos, N. *SCL-90 Analogue*. Baltimore: Johns Hopkins University School of Medicine, 1976.

Hoffmann, N. G., & Overall, P. B. Factor structure of the SCL-90 in a psychiatric population. *Journal of Consulting and Clinical Psychology*, 1978, *46*, 1187–1191.

Jenkins, C. D., Rosenman, R. H., & Zyzanski, S. J. Prediction of clinical coronary heart disease by a test for the coronary prone behavioral pattern. *New England Journal of Medicine*, 1974, *290*, 1271–1275.

Millon, T. *Modern psychopathology*, Philadelphia: W. B. Saunders, 1969.

Millon, T., Green, C., & Meagher, R. *The MBHI manual*. Minneapolis: National Computer Systems, 1982.

Rahe, R. H. *Update to life change researchers*. Unpublished manuscript, 1975.

Rosenman, R. H., Friedman, M., Straus, R., Jenkins, C. D., Zyzanski, S. J., & Wurm, M. A predictive study of coronary heart disease. The Western Collaborative Group Study. *Journal of the American Medical Association*, 1964, *189*, 15.

Rotter, J. B. Generalized expectancies for internal vs. external control of reinforcement. *Psychological Monographs*, 1966, *80* (Whole No. 609), 1–18.

Steptoe, A. *Psychological factors in cardiovascular disorders*. London: Academic Press, 1981.

Willner, A. E. *Conceptual Level Analogy Test (CLAT)*. New York: Cognitive Testing Service, 1971.

Willner, A. E., Rabiner, C. J., Wisoff, B. G., Hartstein, M., Struve, F. A., & Klein, D. F. Analogical reasoning and post operative outcome. *Archives of General Psychiatry*, 1976, *33*, 255–259.

7

Crisis Intervention and Coronary Bypass Patients

Outcome and Research Predictions

FRANKLIN H. FOOTE

Earlier chapters have explained the medical procedures and psychological counseling provided for the coronary bypass patients in this study. The preceding chapter has described the research methods and instruments used in this study as well as the sample itself. This chapter presents statistical analyses of the crisis intervention counseling treatment and nontreatment control conditions.

OVERVIEW OF OUTCOME AND PREDICTIONS

The following are the most important highlights of the outcomes and predictions:

- The crisis intervention counseling reduced the frequency of clinical depression, reports of fears, and suicide thoughts and acts at 3 months after surgery.
- At 3 years after surgery, the crisis intervention counseling reduced depression as measured by reports of a significant family member, usually the patient's wife.
- A three-step precedure involving presence or absence of left main

FRANKLIN H. FOOTE • Department of Psychology, University of Miami, Coral Gables, Florida 33124.

coronary disease, number of bypasses, and the number of items endorsed on the Recent Life Changes Questionnaire appears to predict whether a given patient would benefit from the type of crisis intervention counseling used in this study.

- The crisis intervention counseling and no-treatment control groups had equivalent rates of medical complications during recovery.

A number of other important findings included

- The events associated with bypass surgery appear to have caused changes in patients' coping style.
- General psychopathological symptom severity decreased for all patients 3 months after surgery. Patients also tended to become more hard-driving and competitive.
- Anxious and moody patients were more depressed than other patients. However, they also tended to become less depressed following surgery than they had been before surgery. Passive and conforming patients, on the other hand, tended to become more depressed following surgery.
- Many of the measures of presurgical psychological and medical state affected postsurgical depression in some way. Recent life changes, triple coronary disease, number of bypasses, and history of hypertension were all important moderators of treatment effectiveness.
- An analysis of the impact of combinations of the presurgical variables on postsurgical depression was complex but demonstrated that their combined impact was extremely powerful. Some of the presurgical variables most frequently involved included severity of psychopathological symptoms, recent life changes, age, the presence of triple coronary disease, and the number of bypasses performed.
- There appeared to be a slight effect for pretesting. Administering the psychological assessment battery before surgery tended to increase postsurgical depression. To the extent that this effect occurred, it was equivalent in the crisis intervention counseling and nontreatment control groups.

This chapter is divided into six sections. The first section describes the equivalency of the treatment and control groups prior to surgery. Depression outcome is examined in some detail in the second section and medical outcome is also briefly presented. The third section presents the major predictor equation. Using a simple three-step procedure based on this equation, a physician should be able to predict whether a given individual about to undergo bypass surgery would probably benefit from crisis intervention counseling or whether such treatment would not be beneficial. The fourth section presents some findings which make use of a new psychological instrument, the Millon Behavioral Health Inventory (MBHI). Time and treatment effects on nondepression psychological measures are examined in the fifth section. Finally, the last section examines the possibility of pretesting effects on postsurgical psychological variables and their possible interaction with the treatment. Possible moderating effects of presurgical psychological and medical variables on treatment and postsurgical depression are presented in Appendix A at the end of this chapter. Postsurgical depression as a function of combinations of presurgical psychological and medical variables are detailed in Appendix B. Both concurrent and predictive correlations of the MBHI with other measures are presented in Appendix C.

EQUIVALENCY OF TREATMENT AND CONTROL GROUPS BEFORE SURGERY

T-tests and chi-square statistics were used to determine whether, prior to surgery and any counseling, the treatment and control groups had equivalent levels on a variety of psychological and medical measures. *T*-tests were nonsignificant for the Beck Depression Inventory (BDI), the SCL-90 Global Severity Index (GSI), the SCL-90 Positive Symptom Total (PST), the Rotter Locus of Control Inventory (LOC), the Conceptual Level Analogy Test (CLAT), number endorsed, scaled total, and subjective total for the Recent Life Changes Questionnaire (RLCNE, RLCSCT, and RLCSUT respectively), and the Jenkins Activity Survey (JAS) for both the total score and all three of its subscales, as well as for diastolic and systolic blood pressure and the left-ventricle ejection fraction. The two groups did not differ on any of these measures. Unfortunately, the two groups did differ on the Minnesota Multiphasic Personality Inventory Depression Scale (MMPID), $t\ (65) = -1.98$, $p = .05$. The treatment group had a mean of 63.9 and standard deviation of 13.0, whereas the control group had a mean of 70.3 and standard deviation of 13.1. Because this difference was both statistically significant and in the direction that was the same as that predicted for postsurgical scores, the MMPID scores were dropped from further analyses. Chi-square statistics were calculated for patients' occupation, whether or not they had prior infarction, history of hypertension, family history of heart disease, triple coronary disease, left main coronary disease, previous open-heart surgery, history of angina, number of bypasses, and New York Heart Association Functional Class. Treatment and control groups did not differ on any of these measures. Thus, with the important exception of the MMPID, the treatment and control groups were essentially equivalent.

DEPRESSION OUTCOME

The primary goal of the crisis intervention counseling was the reduction of depression following coronary bypass surgery. The primary measures of depression were the BDI and the Analogue Depression Scale. Since it was hypothesized that the crisis intervention counseling would reduce depression over time and relative to the no-counseling control group, all tests of significance regarding depression outcome were one-tailed. The primary analysis of this data was mixed-design ANOVAs with time (before surgery, 3 months after surgery, and 3 years after surgery) as a repeated measure and Group (treatment versus control) as the between-groups independent variable. Table 1 presents the resulting means, standard deviations, and *F*-ratios for these analyses, which were performed separately for the BDI and Analogue measures. There were no significant effects on the BDI. On the Analogue, however, there was a significant time by Group interaction. As depicted in

Table 1. Means, Standard Deviations, and *F*-Ratios for Depression Measures before and 3 Months and 3 Years after Surgery

Measure	Treatment		Control		*F*-Ratio		
					Group	Time	Group by time
Beck Depression Inventory	$\bar{X}$	*SD*	$\bar{X}$	*SD*			
Before	8.8	4.7	9.4	8.4	0.24	0.68	0.19
3 months after	8.5	5.5	9.9	9.0			
3 years after	7.2	5.4	9.1	10.4			
Analogue Depression Scale[a]							
Before	N/A	N/A	N/A	N/A	0.03	1.50	10.77[b]
3 months after	22.8	31.4	6.3	8.2			
3 years after	11.5	19.3	31.1	31.5			

[a]No Analogue data collected before surgery.
[b]$n = 34$.

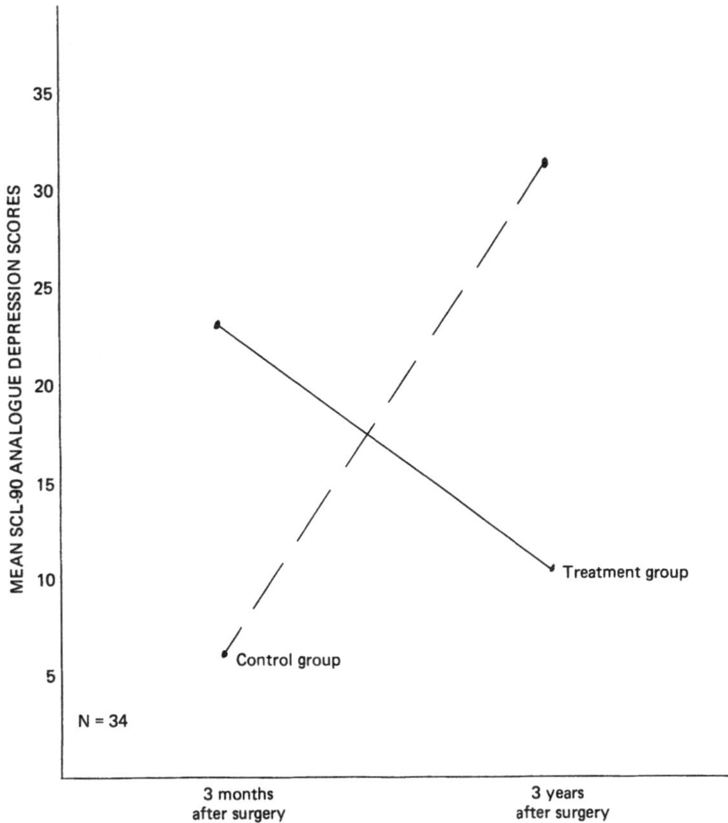

Figure 1. Analogue Depression Scale at 3 months and 3 years.

Figure 1, family members reported a large decrease in depression for patients in the treatment group between the 3-month and the 3-year follow-ups, whereas they reported a large increase for patients in the control group over the same time period.

Since patients with 3-year data represent a small subsample of all patients in the study, a similar pair of analyses was performed on the 3-month data without the 3-year data. This analysis, therefore, benefitted from the increased power and generalizability of the larger sample. Table 2 presents the results of these analyses. Inspection of Table 2 indicates no significant effects for either the BDI or Analogue measure. Careful examination, however, reveals a difference between the BDI means at 3 months after surgery. Unfortunately, the presurgical difference was also large and in the same direction, thus mitigating against the potential significance associated with the postsurgical outcome. To

Table 2. Means, Standard Deviations, and F-Ratios for Depression Measures before and 3 Months after Surgery

Measure	Treatment		Control		F-Ratio		
					Group	Time	Group by time
Beck Depression Inventory	$\bar{X}$	*SD*	$\bar{X}$	*SD*			
Before	8.3	5.4	10.81	8.9	2.92	.19	.16
3 months after	7.6	5.3	10.7	7.9			
Analogue Depression Scale[a]							
Before	N/A	N/A	N/A	N/A	0.19		
3 months after	15.1	21.3	17.7	28.6			

[a]No Analogue data collected before surgery.

explore these differences further, t-tests were calculated before and after surgery. The presurgical difference was not significant, $t(65) = -1.07, p > .10$, whereas the postsurgical difference was significant, $t(94) = -1.86, p = .03$. These analyses cannot be considered as contradicting or superceding the ANOVAs. Nevertheless, they do suggest that there may have been a trend in the predicted direction at 3 months.

Another method of analyzing outcome on depression is to examine the relative frequency with which clinical depression occurs. There is no one generally accepted cutoff for clinical depression on the BDI, but 14 has often been used and was considered reasonable for use in this study. Table 3 displays the numbers of patients who were clinically depressed and not clinically depressed by treatment and control. As shown in Table 3, treatment and control patients had equivalent probabilities of clinical depression presurgery. Three months after surgery, however, control patients were significantly more likely to suffer from clinical depression. At the 3-year mark, the trend was clearly in the same direction but no longer significant. Thus it appears that the crisis intervention counseling reduced the occurrence of clinical depression at 3 months after surgery and that this effect can still be noted 3 years later.

In addition to the primary depression measures, several of the questions asked of the family member at the 3-month follow-up were related to depression. The results of the t-test analyses of these questions are displayed in Table 4. Paralleling the results with the Analogue scale, the family members did not report any differences on depression generally at 3 months after surgery. Inspection of Table 4, however, reveals that family members of patients who had treatment reported that these pa-

Table 3. Frequency Distribution of Clinical Pathology Using Beck Depression Scores

A. Patients before surgery

	Not clinically depressed	Clinically depressed	Total	
Treatment	31	5	36	
Control	24	7	31	χ^2 (1) = 0.86
Total	55	12	67	$p > .10$

B. Patients 3 months after surgery

	Not clinically depressed	Clinically depressed	Total	
Treatment	47	4	51	
Control	35	10	45	χ^2 (1) = 3.97
Total	82	14	96	$p = .05$

C. Patients 3 years after surgery

	Not clinically depressed	Clinically depressed	Total	
Treatment	14	1	15	
Control	13	5	18	χ^2 (1) = 2.45
Total	27	6	33	$p > .10$

tients worried less about death, were less likely to consider suicide, and, in fact, were less likely actually to attempt suicide. The mean score for both groups is small, whereas the amount of variability is high for control patients. These factors are indicative of the possibility that the mean differences are due to a few unusual cases in the control group. Even these differences are important, however, when dealing with an action as extreme as suicide.

The 15 items of the Health Fears and Adjustments following Heart Surgery Questionnaire are all potentially linked to depression. Additionally, since some of the crisis intervention counseling involved communicating objective information, these items are directly related to the treatment strategy. Of the 15 items, two had significant differences between treatment and control groups. Treatment patients reported less fear over irregular heartbeats and noisy valves (mean = 2.3, standard deviation = 0.8) than did control patients (mean = 2.7, standard deviation = 1.2), $t(75.12) = -1.70$, $p = .045$. Additionally, treatment patients said they were less fearful that valves or vessels would be harmed by exertion (mean = 2.7, standard deviation = 0.8), compared with control

Table 4. Means, Standard Deviations, and t Values for Family Member Reports Related to Patient Depression

Question	Treatment $\bar{X}$	Treatment (SD)	Control $\bar{X}$	Control (SD)	t	p
Was patient depressed during out-of-hospital recovery?	1.4	(1.3)	1.5	(1.3)	−0.4	N.S.
Did the patient worry about death during out-of-hospital recovery?	0.4	(0.8)	1.1	(1.4)	−2.8	.007
Did the patient have suicidal thoughts during out-of-hospital recovery?	0.1	(0.6)	0.6	(1.4)	−2.3	.03
Did the patient attempt any suicidal acts during out-of-hospital recovery?	0.0	(0.3)	0.5	(1.4)	−2.2	.03

patients (mean = 3.2, standard deviation = 1.0), $t(76.37) = -2.60$, $p = .005$.

Besides the mean differences between the crisis intervention counseling treatment and the no-treatment control group, correlational differences were also predicted. It was predicted that compared to the control group the crisis intervention counseling group would have a smaller correlation from before to after surgery for depression. This prediction is based on the assumption that the counseling would have most of its impact on cases with relatively greater depression, whereas those who were less depressed would not be helped so much simply because they had less need for help. This prediction was supported. The treatment group had a Pearson Product Moment Correlation of .46 from before surgery to 3-month follow-up on the BDI and the control group had a correlation of .69. Using an r to z transformation, this difference was marginally significant at the less than .10 level. For before surgery to 3-year follow-up, the correlation was .46 for the treatment group and .83 for the control group. This difference was significant at the .05 level. Thus the hypothesis that the impact of treatment would reduce the correlation between pre- and postsurgical depression was supported.

Additional analyses examined the effects of individual measures of psychological and medical variables on later depression and on the effectiveness of treatment. These results are presented in detail in Appendix A at the end of this chapter. Briefly, many variables had some sort of impact. Several effects were particularly interesting. Treatment was particularly helpful in reducing depression for people with many recent life changes, with triple coronary disease, or having many bypasses. Patients with a history of hypertension did relatively well at 3 months but poorly at 3 years. Patients low in Cooperative Personality Style or high in Forceful Personality Style on the MBHI did better in the treatment condition.

Another set of analyses discovered which combination of variables predicted later depression. These analyses also contribute to an understanding of postsurgical depression. The complex details of the results of these analyses are presented in Appendix B. The most striking finding is that the impact of presurgical psychological and medical state is extremely powerful. Prominent variables in these analyses included number of recent life changes endorsed, SCL-90 Global Severity Index, age, presence of triple coronary disease, and number of bypasses performed.

The primary aim of the crisis intervention counseling was the reduction of depression and associated psychopathology. Nevertheless, examination of the effect of the treatment on medical outcome is obviously essential. Although complete medical data is not yet available, several possible catastrophic complications were analyzed for almost half the patients in the study. These complications were renewed angina, myocardial infarction, other serious disease, or death. The occurrence of any one of these events was rare, however. Therefore, patients were classified into two groups: those who experienced any one or more of these complications and those who experienced none of them. The counseling treatment and no-treatment control groups had equivalent rates of this composite of complications. At 3 months after surgery 16% of the treatment patients had one or more of these complications, whereas 20% of the control group patients had them. At 3 years after surgery the composite rates of these complications were exactly equal, with one-third of the patients in both treatment and control groups experiencing them.

In summary, the crisis intervention counseling reduced depression at 3 months in terms of clinical depression and as measured by the reports of fears and of suicidal thoughts and acts. At 3 years, depression as measured by the reports of a significant family member on the Analogue scale was significantly reduced. Additionally, the treatment reduced the correlation between pre- and postsurgical depression, especially at 3 years. Finally, the treatment and control groups had equivalent rates of medical complications at both 3 months and 3 years after surgery.

PREDICTING THE BENEFIT OF CRISIS INTERVENTION COUNSELING

While the outcome results suggest that the treatment for depression suffered by bypass patients was generally successful, it is not likely to be necessary or even helpful for all such patients. This section will describe a multiple regression equation that is simple to apply and can predict whether or not a given individual would benefit from the type of crisis intervention counseling treatment used in this study. Although the equations in Appendix B might be helpful in this regard, they were designed to be more general and descriptive. The equation in this section is designed for the express purpose of prediction. The derivation of the equation and the independent variables included in it are described first. After this analysis, the application of the equation is presented.

Given its predictive purpose, only those variables that might be helpful for this purpose were included in the pool of potential independent variables. These variables include the group variable itself and the interactional variables (Group by Intake Measure) described in more detail in Appendix B. Thus, these variables did not have to compete with all the main effect intake measures as they did in the analysis in Appendix B. Additionally, a small cross-validation study was conducted.

In order to have as many patients as possible included in the analysis, BDI at 3 months was chosen as the dependent variable. Criteria for independent variable inclusion in the equation was an *F*-ratio for inclusion with a significance level less than .10. For a variable to be excluded once it was included, an *F*-ratio for exclusion with significance level less than .125 was required.

Table 5 presents the results of the analysis. A highly significant multiple *R* was obtained with three significant independent variables (plus a constant of intersection for the unstandardized coefficients). These variables and their direction were as follows: (1) Group by RLCNE, being in the control group and having many recent life changes presurgically is associated with greater postsurgical depression; (2) Group by Number of Bypasses, being in the control group and having more bypasses is associated with greater postsurgical depression;

Table 5. Regression Analysis for Predicting the Usefulness of Treatment

Equation statistics

(Beck Depression Inventory at 3 years is dependent variable)
$R = .70$, $R^2 = .49$, $R^2_{Adj.} = .44$, $F(3,29) = 9.25$, $p = .0002$

Independent variable statistics

Variable	Unstandardized regression coefficient	Standard regression coefficient	Standard error of standard regression coefficient	F	p
Group by Recent Life Changes Number Endorsed	.46	.62	.14	19.73	.0001
Group by Number of Bypasses	1.01	.41	.14	8.38	.007
Group by Left Main Coronary Disease	−2.97	−.39	.15	6.88	.01
Constant	6.34	—	—	5.44	.03

Cross-validation sample
$r = .56$, $r^2 = .31$, $p = .04$

(3) Group by Left Main Coronary Disease, being in treatment and having left main coronary disease are associated with greater depression. This equation explained almost half the variance in postsurgical BDI depression among those patients who were used to derive it. A total of 11 patients had data on all five variables involved with this equation but were missing data on one or more of the other potential independent variables which did not enter into the equation. These 11 patients, therefore, could be used as a cross-validation sample (even though, because of differences in missing data, they might be considered to come from a different population). As presented in Table 5, the Pearson Product Moment Correlation between their actual BDI score at 3 months and the predicted BDI score from the equation was .56 ($p = .04$). Thus, even in this cross-validation sample, the equation accounted for more than 30% of the variance in postsurgical depression.

Using This Analysis: Tough Way

The most straightforward method of using this analysis would be to use the unstandardized regression coefficients as in the following:

$$D = (.46 \times G \times R) + (1.01 \times G \times B) - (2.97 \times G \times L) + 6.34$$

where D = predicted postsurgical BDI score; G = group (treatment = 1, no treatment = 2); R = number endorsed on Rahe's Recent Life Change Scale; B = number of bypasses; L = left coronary disease (present = 1, absent = 2).

This equation would be calculated twice, once with $G = 1$ and once with $G = 2$. If the score is lower with $G = 1$, the patient would probably benefit from crisis intervention counseling. If the score is higher with $G = 1$, then the patient would probably be better off without the crisis intervention counseling. Note that to use this equation all four terms must be known. Do not use estimates.

To illustrate the use of this equation, consider two hypothetical patients. Joe Brown had endorsed two items on Rahe's Recent Life Change Scale, was to have four bypasses, but did not have left main coronary disease. If $G = 1$ (i.e., treatment) then the above equation is

$$D = (.46 \times 1 \times 2) + (1.01 \times 1 \times 4) - (2.97 \times 1 \times 1) + 6.34 = 8.33.$$

If $G = 2$ (i.e., no treatment) then the equation becomes

$$D = (.46 \times 2 \times 2) + (1.01 \times 2 \times 4) - (2.97 \times 2 \times 1) + 6.34 = 10.32.$$

Since the equation with $G = 1$ produces a lower sum, Mr. Brown would probably benefit from crisis intervention counseling.

Now consider the case of George Beckford. He also had endorsed two items on the Rahe's Recent Life Change Scale, he was to have only two bypasses, but he did have left main coronary disease. For Mr. Beckford the equation with $G = 1$ is

$$D = (.46 \times 1 \times 2) + (1.01 \times 1 \times 2) - (2.97 \times 2 \times 2) + 6.34 = 3.34.$$

If $G = 2$ then the equation for Mr. Beckford is

$$D = (.46 \times 2 \times 2) + (1.01 \times 2 \times 2) - (2.97 \times 2 \times 2) + 6.34 = .34.$$

Since the equation with $G = 2$ produces a lower sum, Mr. Beckford would probably be less depressed if he did not receive crisis intervention counseling.

Using This Analysis: Easy Way

Careful examination of Table 5 reveals two interesting facts: First, according to the analysis, the only time depression will be lower without treatment is if the patient has left main coronary disease. Second, if the patient does have left main coronary disease, the other two variables are

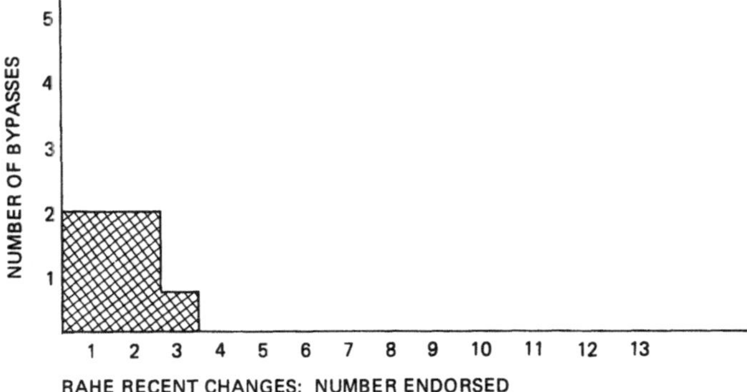

Figure 2. Recent life change by number of bypasses.

a simple function. This function is pictured in Figure 2. A simple three-step procedure can be followed. Table 6 presents these three steps.

Now reconsider the hypothetical patients Mr. Brown and Mr. Beckford. Step 1 of Table 6 states that if no left main coronary disease is present, the patient should receive crisis intervention counseling. Since Mr. Brown does not have this disease, he should receive counseling. No other calculations are necessary! Since Mr. Beckford does have left main coronary disease proceed to Step 2. If Mr. Beckford needed three or more bypasses, then he too should have counseling. However, since Mr. Beckford was having only two bypasses, proceed to Step 3.

Table 6. Three-Step Procedure for Deciding the Benefit of Crisis Intervention Counseling to a Patient

1. If the patient does *not* have left main coronary disease, then crisis intervention counseling should accompany his coronary bypass surgery regardless of the other factors. If the patient does have left main coronary disease, go to Step 2.
2. If the patient is to have three or more bypasses, then he should receive crisis intervention counseling regardless of other factors. If the patient is to have one or two bypasses, go to Step 3.
3. If the patient does have left main coronary disease and will have two or fewer bypasses, then have the patient take Rahe's Recent Life Changes inventory (note that it is important to use this instrument and *not* some other measure of recent life change). Check the number endorsed on the Recent Life Changes and the number of bypasses against Figure 2. If the patient's number of bypasses and number of recent life changes falls within the shaded area, he probably should *not* receive crisis intervention counseling. If the patient's number of bypasses and number of recent life changes fall outside of the shaded area, he probably should receive crisis intervention counseling.

For Step 3 examine Figure 2. The intersection of two bypasses and two life changes is in the shaded area on Figure 2; therefore, Mr. Beckford should not need or receive crisis intervention counseling.

This analysis, of course, applies only to male patients about to receive coronary bypass surgery considering crisis intervention counseling of the type used in this study described in Chapter 5. To the extent that circumstances differ from those described for the patients in this study, the results of this analysis should be interpreted with caution. To the extent that circumstances are similar, however, use of the three-step procedure given in Table 6 will result in the best estimate of the way to minimize a patient's postsurgical depression.

It is also important to remember that this equation is based on a *post hoc* analysis of data and has not been tested prospectively on a group of coronary bypass patients. Until this is done, it is suggested that the reader use this equation with this caution in mind.

MILLON BEHAVIORAL HEALTH INVENTORY AND DEPRESSION FOLLOWING CORONARY BYPASS SURGERY

In a master's thesis using an early subsample of the present study Levine (1980) hypothesized that patients with one of the three major coping styles as measured by the eight Personality Style scales of the MBHI would respond differently when confronting coronary bypass surgery. The first style, labeled "Anxious and Moody," consists of patients with elevated Inhibited, Sensitive, or Forceful Personality Style Scales. The second type of coping style, labeled "Passive and Conforming," consists of patients with elevated Introversive, Cooperative, or Respectful Personality Style Scales. The third style is labeled "Confident and Sociable" and includes those with elevated Confident or Sociable Scales.

Before examining the effects of these three major coping styles on depression and the efficacy of the crisis intervention counseling, we will demonstrate their stability across the three assessments in this study. Table 7 presents this stability information as frequency cross-tabulations. Part A displays the stability from before surgery to 3 months after surgery. The chi-square for this relationship is nonsignificant ($\chi^2(4) = 6.23$, $p < .10$), indicating that the relationship between coping styles before and after surgery was not reliable. Examination of Table 7(A) suggests that the Confident and Sociable coping style showed the most instability. In particular, there was a great deal of crossover from Confi-

Table 7. Millon Coping Style Stability over Time

A. Before to after surgery

	3 months after surgery			
	Anxious and Moody	Passive and Conforming	Confident and Sociable	Total
Before surgery				
Anxious and Moody	6	3	2	11
	54.5%	27.3%	18.2%	100%
Passive and Conforming	4	10	2	16
	25.0%	62.5%	12.5%	100%
Confident and Sociable	3	6	5	14
	21.4%	42.9%	33.7%	100%
Total	13	19	9	41

χ^2 (4) = 6.23, $p > .10$

B. 3 months to 3 years after surgery

	3 years after surgery			
	Anxious and Moody	Passive and Conforming	Confident and Sociable	Total
3 months after surgery				
Anxious and Moody	7	1	2	10
	70.0%	10.0%	20.0%	100%
Passive and Conforming	2	13	1	16
	12.5%	81.3%	6.3%	100%
Confident and Sociable	1	0	4	5
	20.0%	—	80.0%	100%
Total	10	14	7	31

χ^2 (4) = 25.41, $p < .00001$

dent and Sociable to Passive and Conforming. In Part B of Table 7 the stability from 3 months to 3 years is displayed. The chi-square for this table is highly significant, $\chi^2(4) = 25.41$, $p < .0001$. Thus the coping styles were highly stable during this time period. Although the time period was much shorter for Table 7(A) than for Table 7(B), the stability was much greater in Table 7(B) than in Table 7(A). Apparently either the stress experienced the day before surgery (when patients were administered the psychological instruments) or during postsurgical recovery lead to changes in major coping style. In more normal circumstances, however, coping style was highly stable even across a 2½-year time period.

The effect of presurgical coping style on depression and the efficacy of the crisis intervention counseling is presented in Table 8. No results for 3-year data are presented in this table because of insufficient numbers in some of the six between groups (i.e., treatment versus control by three types of coping style) for the 3-year sample. The three coping styles had no effect on the Analogue Depression measure at 3 months after surgery. However, on the BDI the main effect for coping style and the interaction for coping style by time were both significant. Anxious and moody patients tended to have higher depression both before and after surgery. In addition, whereas anxious and moody patients tended to become less depressed following surgery, passive and conforming patients tended to become more depressed after surgery and confident and sociable patients showed little consistent change.

Levine (1980) also suggested that regardless of coping style, the combination of Premorbid Pessimism, Future Despair, and Social Alienation should be predictive of postsurgical depression. The mean of these three scales for each patient comprises the combination score. Table 9 presents the relation between this combination score and the depression measure. Not surprisingly, as shown in Table 9, the patients with the Anxious and Moody coping style and higher scores on this combination of the three MBHI scales than did patients with other coping styles. Also, as shown in Table 9, across all three coping styles, the correlation between the combination of the three MBHI scales presurgically and the BDI and Analogue Depression Scale at all three assessments was statistically significant although stronger for the BDI than the Analogue scale. Among the three coping styles separately, the relationship was less uniform. Only for confident and sociable patients was there a failure to find any such significant relationship, however. Thus the data were consistent with the notion that the combination of Premorbid Pessimism, Future Despair, and Social Alienation is predictive of later depression but only for those patients with Anxious and Moody or Passive and Conforming coping styles.

For the information of those readers for whom the MBHI is a new instrument, the correlation between each of the presurgical MBHI scales and the other standard psychological and medical instruments used in this study are presented in Appendix C at the end of this chapter. Both concurrent and predictive correlations are given. Thus careful examination of Appendix C will give some additional empirical meaning to the MBHI scales.

Table 8. Millon Coping Style Effect on Depression

	Mean (standard deviation)						F-Ratio[a]						
	Treatment			Control									
	Anxious and Moody	Passive and Conforming	Confident and Sociable	Anxious and Moody	Passive and Conforming	Confident and Sociable	G	C	GC	T	TG	TC	TGC
Beck													
Before surgery	14.7	8.3	6.7	19.0	5.9	6.8	0.8	6.2**	0.4	0.4	0.1	4.5*	0.3
	(8.3)	(8.2)	(6.3)	(11.7)	(2.5)	(4.2)							
3 months after surgery	11.0	9.1	5.2	14.1	9.5	6.8							
	(9.5)	(6.7)	(3.9)	(12.1)	(5.5)	(2.9)							
Analogue Depression													
3 months after surgery	5.0	20.3	11.0	22.0	32.0	2.8	0.1	2.3	0.8				
	(8.7)	(25.9)	(20.0)	(26.4)	(39.0)	(5.1)							

[a]G = Group effect (treatment versus control), C = Coping Style effect (Anxious and Moody versus Passive and Conforming versus Confident and Sociable), T = time.
$*p < .05$. $**p < .01$.

Table 9. Relation of Depression and Mean of Presurgical MBHI Premorbid Pessimism, Future Despair, and Social Alienation by Three Major Personality Coping Styles

Group	Mean of presurgical MBHI Premorbid Pessimism, Future Despair, and Social Alienation		Mean of presurgical MBHI Premorbid Pessimism, Future Despair, and Social Alienation				
	Mean	STD	r with before surgery BDI	r with 3 months after surgery BDI	r with 3 years after surgery BDI	r with 3 months after surgery Analogue	r with 3 years after surgery Analogue
All patients	40.6	25.0	.630***	.605***	.727***	.375*	.408†
Anxious and Moody	64.3^a	21.9	.681*	.667**	.786	.373	.608
Passive and Conforming	35.3^b	19.3	.339	.578*	.665*	.722**	.181
Confident and Sociable	24.2^b	17.0	.136	.041	.479	−.220	.512

^{a,b}Means with differing superscripts differ at better than .05 level on the Duncan Multiple Range test. Overall $F(344) = 16.0$, $p < .001$.
$^{\dagger}p < .10$. $^{*}p < .05$. $^{**}p < .01$. $^{***}p < .001$.

TIME AND TREATMENT EFFECTS ON PSYCHOLOGICAL MEASURES OTHER THAN DEPRESSION

In addition to the depression measures, several other psychological variables were measured before surgery and 3 months and/or 3 years after surgery. The effects of treatment and time (including the effects of surgery) were assessed with mixed design ANOVAs of Group as a between-groups independent measure and time before surgery, 3 months, and/or 3 years after surgery) as a repeated measure.

Table 10 presents results that include a measurement at 3 years after surgery. One set of significant effects involved the Jenkins Activity Survey (JAS) as the dependent measure. There was a Group effect on the total JAS score and the Speed and Impatience subscale. On both these scales, the control group tended to have higher scores (i.e., more like Type B's). Additionally, the Jenkins Hard Driving Competitive Subscale had a main effect for time with both the treatment and the control group becoming more like Type A's at 3 years.

Another set of effects involved the MBHI scales and are also displayed in Table 10. The Introversive Personality Style and Social Alienation scales both showed a significant time effect such that there was an increase at 3 years. There was also a main effect for time on the Recent Stress scale such that there was a decrease at 3 years. In addition to the main effect for time there was also a marginally significant interaction effect for the Recent Stress scale. This marginal effect appears to be due to presurgical differences on this scale such that treatment patients had a lower score than control patients.

Table 11 presents results of the mixed-design ANOVAs with time measured before surgery and 3 months after surgery. The two SCL-90 Summary Scales, GSI and PST, both had significant time effects. For both scales, both treatment and control groups had less symptomology after surgery. Similarly, the Recent Life Changes scaled total had a marginally significant time effect such that both treatment and control groups experienced less stress after surgery than before surgery. The fact that the Number Endorsed and Subjective Total scales did not show a time effect is interesting since the time span for Recent Life Changes before surgery covered 6 months whereas that after surgery covered only 3 months.

Also as displayed in Table 11, on the MBHI scales there were several main effects but they were only marginally significant. Introversive Personality Style had a marginal tendency to increase from before to after surgery and Somatic Anxiety had a marginal tendency to decrease. Additionally, patients in the treatment condition had a marginal tenden-

Table 10. Means, Standard Deviations, and *F*-Ratios for Nondepression Measures before and 3 Months and 3 Years after Surgery

Measure	Treatment		Control		*F*-Ratio		
	$\bar{X}$	*SD*	$\bar{X}$	*SD*	Group	Time	Group by time
SCL-90 Global Severity Index							
Before	59.8	(11.2)	60.3	(9.4)	0.06	0.73	0.53
3 months	59.5	(11.4)	56.9	(8.6)			
3 years	59.2	(11.0)	58.7	(12.6)			
SCL-90 Positive Symptom Total							
Before	56.8	(12.7)	59.5	(8.2)	0.01	0.09	2.29
3 months	59.5	(10.1)	55.8	(6.9)			
3 years	58.2	(9.8)	58.3	(12.1)			
Locus of Control							
Before	7.6	(4.7)	7.0	(3.6)	0.13	0.95	1.67
3 months	5.8	(3.6)	6.8	(3.7)			
3 years	8.2	(4.7)	6.3	(4.1)			
Jenkins Total Type A Standard Score							
Before	25.8	(11.0)	32.7	(8.8)	5.75*	1.67	0.04
3 months	N/A	N/A	N/A	N/A			
3 years	24.2	(7.4)	30.6	(8.4)			
Jenkins Type A Speed and Impatience Subscale							
Before	24.2	(10.7)	33.0	(9.4)	5.83*	0.06	0.69
3 months	N/A	N/A	N/A	N/A			
3 years	25.6	(9.7)	32.2	(9.9)			
Jenkins Type A Job Involvement Scale							
Before	21.9	(10.0)	22.3	(9.5)	0.02	0.63	0.15
3 months	N/A	N/A	N/A	N/A			
3 years	24.1	(7.8)	23.1	(8.5)			
Jenkins Type A Hard Driving Competitiveness Subscale							
Before	26.6	(14.3)	31.3	(7.4)	0.52	4.85*	1.53
3 months	N/A	N/A	N/A	N/A			
3 years	24.4	(7.6)	23.3	(8.3)			
Introversive							
Before	54.2	(25.1)	44.0	(23.8)	0.46	4.54*	0.74
3 months	55.6	(22.8)	46.2	(23.9)			
3 years	60.9	(30.6)	59.3	(27.9)			

(*continued*)

Table 10. (*Continued*)

Measure	Treatment $\bar{X}$	Treatment *SD*	Control $\bar{X}$	Control *SD*	*F*-Ratio Group	*F*-Ratio Time	*F*-Ratio Group by time
Inhibited							
Before	37.2	(28.7)	36.8	(24.8)	0.00	0.60	0.02
3 months	38.4	(33.3)	37.0	(30.4)			
3 years	40.6	(32.8)	40.3	(29.1)			
Cooperative							
Before	63.9	(26.6)	44.3	(27.4)	1.09	1.63	1.16
3 months	55.7	(29.8)	46.8	(22.3)			
3 years	46.6	(27.8)	42.8	(29.4)			
Sociable							
Before	53.2	(29.6)	53.7	(20.6)	0.05	0.38	0.09
3 months	53.8	(31.8)	57.2	(19.9)			
3 years	50.6	(30.4)	53.8	(23.8)			
Confident							
Before	39.0	(22.3)	55.9	(25.2)	1.99	0.68	0.72
3 months	41.0	(27.8)	55.2	(20.4)			
3 years	47.7	(24.6)	55.6	(19.3)			
Forceful							
Before	27.3	(26.8)	47.7	(27.3)	1.88	0.99	0.59
3 months	34.0	(30.7)	49.3	(22.5)			
3 years	39.9	(30.4)	49.3	(31.2)			
Respectful							
Before	47.1	(23.9)	63.4	(13.4)	2.59	0.60	1.24
3 months	46.6	(26.5)	60.7	(13.9)			
3 years	48.4	(21.8)	54.9	(17.1)			
Sensitive							
Before	38.0	(27.5)	43.3	(22.7)	0.04	0.26	0.28
3 months	38.3	(31.2)	40.6	(25.8)			
3 years	38.1	(29.7)	37.4	(31.3)			
Chronic Tension							
Before	30.1	(29.2)	55.8	(17.5)	4.04†	0.29	0.90
3 months	35.4	(32.1)	52.9	(19.1)			
3 years	33.7	(23.4)	48.3	(25.8)			
Recent Stress							
Before	30.2	(21.1)	55.0	(29.8)	1.30	6.15**	2.86†
3 months	47.9	(21.8)	47.8	(24.6)			
3 years	25.6	(16.3)	34.3	(34.2)			
Premorbid							
Before	32.8	(25.8)	41.0	(29.7)	0.27	0.20	1.01
3 months	39.0	(28.9)	37.7	(25.3)			
3 years	30.4	(21.4)	40.7	(31.2)			
Future Despair							
Before	41.9	(29.5)	41.1	(27.7)	0.01	1.69	0.01

(*continued*)

Table 10. (*Continued*)

Measure	Treatment		Control		*F*-Ratio		
	$\bar{X}$	*SD*	$\bar{X}$	*SD*	Group	Time	Group by time
3 months	41.4	(28.2)	39.8	(28.3)			
3 years	47.9	(20.6)	46.8	(31.4)			
Social Alienation							
Before	29.4	(26.7)	25.9	(22.7)	0.29	7.27**	0.55
3 months	38.7	(29.4)	28.5	(24.8)			
3 years	43.8	(23.6)	41.0	(25.6)			
Somatic Anxiety							
Before	47.2	(26.3)	49.9	(26.0)	0.13	1.26	0.74
3 months	43.4	(24.1)	42.3	(25.8)			
3 years	38.0	(17.0)	47.1	(27.6)			
Allergic Inclination							
Before	49.9	(27.5)	60.0	(21.4)	0.98	1.07	0.01
3 months	48.3	(29.6)	57.8	(26.3)			
3 years	53.1	(23.2)	63.8	(23.1)			
Gastrointestinal Susceptibility							
Before	55.1	(20.0)	61.5	(22.8)	0.10	0.06	0.82
3 months	60.6	(19.1)	58.3	(22.5)			
3 years	56.7	(13.7)	60.4	(23.2)			
Cardiovascular Tendency							
Before	47.6	(23.6)	65.3	(14.3)	2.20	0.05	1.11
3 months	50.1	(27.3)	60.8	(16.2)			
3 years	51.9	(21.1)	59.3	(21.0)			
Pain Proneness							
Before	30.2	(29.5)	31.5	(33.9)	0.09	0.99	0.53
3 months	27.2	(15.1)	23.8	(32.3)			
3 years	30.0	(14.4)	40.6	(31.4)			
Life Threat Reactivity							
Before	47.1	(27.1)	51.8	(27.3)	0.36	0.22	0.22
3 months	46.2	(25.4)	50.1	(22.9)			
3 years	46.4	(14.8)	55.5	(27.4)			
Emotional Vulnerability							
Before	18.9	(23.1)	32.1	(28.0)	0.32	0.23	0.77
3 months	21.7	(20.9)	23.8	(23.8)			
3 years	21.7	(17.7)	22.1	(32.8)			

$^{\dagger}p < .10$. $^{*}p < .05$. $^{**}p < .01$.

Table 11. Means, Standard Deviations, and *F*-Ratios for Nondepression Measures before and 3 Months after Surgery

Measure	Treatment		Control		*F*-Ratio		
	$\bar{X}$	*SD*	$\bar{X}$	*SD*	Group	Time	Group by time
SCL-90 Global Severity Index							
Before	58.3	(11.0)	59.8	(10.6)	0.63	4.64*	0.11
3 months	55.4	(10.4)	57.7	(10.3)			
SCL-90 Positive Symptom Total							
Before	56.6	(10.6)	59.1	(9.5)	0.71	4.57*	0.44
3 months	54.9	(9.8)	56.0	(8.4)			
Locus of Control							
Before	6.8	(4.1)	8.1	(3.8)	2.67	0.81	1.18
3 months	5.9	(3.5)	8.2	(4.8)			
Recent Life Changes Number Endorsed							
Before	5.0	(3.7)	6.4	(5.7)	0.53	1.04	0.50
3 months	4.8	(4.1)	5.2	(4.1)			
Recent Life Changes Scales Total							
Before	138.7	(78.1)	158.6	(133.1)	0.20	3.16†	0.21
3 months	116.6	(79.7)	121.2	(89.3)			
Recent Life Changes Subjective Total							
Before	174.2	(132.7)	247.3	(223.8)	0.41	0.81	0.51
3 months	238.4	(179.3)	254.7	(250.3)			
Concept Level Analogy Test							
Before	8.9	(5.7)	9.5	(7.1)	0.24	2.11	0.07
3 months	9.8	(5.2)	10.8	(4.5)			
Introversive							
Before	53.4	(28.0)	41.8	(24.3)	2.16	3.33†	0.02
3 months	57.9	(28.5)	47.2	(22.7)			
Inhibited							
Before	38.3	(25.7)	39.0	(27.0)	0.01	0.00	0.00
3 months	38.2	(30.0)	38.8	(26.9)			
Cooperative							
Before	56.2	(27.3)	43.9	(26.6)	2.05	0.00	0.12
3 months	54.9	(28.5)	45.1	(26.7)			

(continued)

Table 11. (*Continued*)

Measure	Treatment		Control		*F*-Ratio		
	$\bar{X}$	*SD*	$\bar{X}$	*SD*	Group	Time	Group by time
Sociable							
Before	51.7	(26.0)	55.5	(23.5)	0.12	0.00	0.17
3 months	52.9	(28.6)	54.1	(22.6)			
Confident							
Before	47.4	(25.7)	53.1	(23.1)	1.21	0.00	0.24
3 months	45.8	(28.5)	54.7	(17.5)			
Forceful							
Before	37.7	(29.7)	51.0	(26.8)	3.31†	0.40	0.10
3 months	34.3	(26.7)	49.9	(27.4)			
Respectful							
Before	53.0	(20.5)	61.4	(15.8)	1.94	0.14	0.16
3 months	53.0	(21.5)	59.8	(16.9)			
Sensitive							
Before	40.4	(27.1)	45.7	(28.0)	0.38	1.13	0.00
3 months	38.9	(29.5)	44.1	(28.2)			
Chronic Tension							
Before	36.7	(30.8)	53.6	(19.9)	4.01†	0.92	0.19
3 months	35.3	(31.1)	49.9	(24.5)			
Recent Stress							
Before	34.8	(24.4)	52.0	(28.8)	1.94	2.25	5.06†
3 months	46.5	(23.7)	49.7	(23.5)			
Premorbid Pessimism							
Before	37.4	(23.0)	43.1	(28.3)	0.16	0.38	0.81
3 months	41.8	(27.6)	42.3	(24.9)			
Future Despair							
Before	44.8	(26.9)	46.0	(28.4)	0.00	0.29	0.10
3 months	44.1	(27.1)	43.5	(28.9)			
Social Alienation							
Before	29.6	(24.7)	27.9	(25.9)	0.59	1.99	1.79
3 months	37.9	(26.1)	28.1	(25.4)			
Somatic Anxiety							
Before	51.4	(25.3)	52.3	(23.3)	0.07	3.23†	0.12
3 months	45.6	(25.3)	48.4	(24.1)			
Allergic Inclination							
Before	53.3	(24.7)	61.8	(22.6)	1.24	1.71	0.02
3 months	50.4	(26.4)	58.3	(25.4)			

(*continued*)

Table 11. (*Continued*)

Measure	Treatment		Control		*F*-Ratio		
	$\bar{X}$	*SD*	$\bar{X}$	*SD*	Group	Time	Group by time
Gastrointestinal Susceptibility							
Before	58.3	(17.8)	64.8	(20.4)	0.50	0.21	1.13
3 months	61.8	(18.6)	63.4	(20.9)			
Cardiovascular Tendency							
Before	50.0	(21.5)	62.6	(19.4)	3.06†	0.13	0.86
3 months	51.3	(23.4)	59.6	(17.4)			
Pain Proneness							
Before	30.8	(23.8)	38.1	(33.7)	0.06	1.12	1.25
3 months	31.1	(21.4)	27.7	(31.3)			
Life Threat Reactivity							
Before	52.2	(24.2)	56.5	(25.5)	0.22	0.24	0.15
3 months	51.9	(24.6)	54.3	(22.8)			
Emotional Vulnerability							
Before	26.4	(23.7)	29.3	(27.2)	0.14	1.60	0.01
3 months	21.9	(24.2)	24.1	(23.9)			

†$p < .10$. *$p < .05$.

cy to have lower Forceful Personality Style and lower Cardiovascular Tendency than did patients in the control condition. Finally, there was a significant interaction on Recent Stress, which increased from before to after surgery among treatment patients but decreased among control patients.

Thus, across both time and Group there were relatively few effects. Psychopathological symptoms severity was reduced at 3 months but showed no effect at 3 years. Patients also became slightly more Type A after surgery but only on the Hard Driving–Competitiveness subscale.

PRETESTING EFFECTS

Recently there has been some concern that the assessment process itself might alter later outcome (e.g., Mahoney, 1978). In order to investigate the possibility that pretesting might have affected depression postsurgically, 24 patients were deliberately not given pretests. At 3 months after surgery those who received pretesting had a mean BDI score of 9.2 (and standard deviation of 6.8) whereas those who did not receive pretesting had a mean BDI score of 5.9 (and standard deviation of 4.4). This difference was marginally significant, $F(2,94) = 2.92$, $p = .06$. There was, however, no significant interaction between pretesting and treatment, $F(2,95) = 1.1$, $p > .10$. For the Analogue scale the group that had been pretested had a mean of 17.6 (and standard deviation of 26.9) and the group that had not been pretested had a mean of 15.2 (and a standard deviation of 17.5). Neither this difference nor the interaction between pretesting and treatment was significant, $F(2,68) < 1$ for both.

Because of the difference between the pretesting and no-pretesting groups on the BDI, other psychological variables were examined for pretesting effects. There were no pretesting effects on any of these other variables, which included the LOC, the CLAT, the GSI, the PST, and the three Recent Life Change Questionnaire scales.

Thus the analyses of potential pretesting effects were somewhat mixed. On the one hand, the effect that was shown was only marginally significant and occurred on only one of nine variables examined. On the other hand, the difference that did occur appeared to be a sizeable one and involved a variable of central interest. The fact that, as described in Chapter 6, the patients could not be randomly assigned to pretesting and no-pretesting groups further complicates the interpretation of these results. The best general conclusion is that pretesting may have contributed to some increased depression in some cases. To the extent that pretesting did have such an effect, it did not interact with treatment. That is, the degree of increase in depression due to pretesting appeared to be equivalent in both the crisis intervention counseling group and the no-treatment control group.

Thus, with this population of white Anglo-American males around 60 years of age, the crisis intervention counseling treatment was generally helpful. These and other results will be discussed in subsequent chapters.

REFERENCES

Levine, R. *The impact of personality style upon emotional distress, morale and return to work in two groups of coronary bypass surgery patients.* Unpublished master's thesis, University of Miami, 1980.

Mahoney, M. J. Experimental methods and outcome evaluation. *Journal of Consulting and Clinical Psychology*, 1978, *46* (4), 660–672.

A

Presurgical Interaction with Depression Outcome

This study of crisis intervention counseling on depression in coronary bypass patients examined a number of psychological and medical variables. From the study's inception, the investigators were well aware that any one of these variables might interact with the counseling treatment by facilitating it, hindering it, or even negating it. In order to assess the impact of each of these psychological and medical variables on treatment and later depression, four sets of ANOVAs was calculated. These ANOVAs are similar to but somewhat more complex than those discussed earlier in the chapter (cf. pp. 94–96). Patients were divided into two groups on each presurgical measure. For the Jenkins, patients were divided in the standard manner into Type A's and Type B's. On the medical history variables they were divided into "yes" "no" groups. On the continuous measures (e.g., LOC, Ejection Fraction) they were divided at the median of all scores. These two groups were factorially crossed with the treatment and control groups yielding four groups. These four groups were then examined across time. Table A-1 presents the means, standard deviations, and *F*-ratios for the results of the 2 × 2 × 3 (i.e., treatment versus control groups, two levels of each intake measure, and time—before surgery, 3 months after surgery, and 3 years after surgery—as a repeated measure) ANOVA with Beck Depression Inventory as the dependent measure. Inspection of Table A-1 reveals relatively few effects. High general symptomatology as measured by both the GSI and the PST on the SCL-90-R resulted in significantly higher depression at all three times in both groups. A trend for externality on the Locus of Control Scale to have greater depression was only marginally significant. Many of the MBHI scales had significant main effects. These included Inhibited Personality Style, Sociable Personality Style, Recent Stress, Premorbid Pessimism, Future Despair, Social Alienation, Somatic Anxiety, Gastrointestinal Susceptibility, and Life Threat Reactivity. In addition, Sensitive Personality Style, Chronic Tension, Allergic Inclination, Cardiovascular Tendency, and Pain Proneness had marginally significant main effects. For all these MBHI scales except the Sociable Personality Style a low presurgical score was associated with a low BDI score at all three assessments in both treatment and control groups. For the Sociable Personality Style a high score was associated with a low BDI score. The only variable to have a significant

Table A-1. Group by Various Intake Measures for the Beck Depression Inventory before and 3 Months and 3 Years after Surgery

	Treatment				Control				F-Ratio[a]						
	Low IM		High IM		Low IM		High IM								
Intake measure	$\bar{X}$	SD	$\bar{X}$	SD	$\bar{X}$	SD	$\bar{X}$	SD	G	IM	GIM	T	TG	TIM	TGIM
Locus of Control															
Before	6.2	(4.5)	10.4	(4.3)	6.3	(4.7)	14.4	(11.0)	0.55	3.20†	0.28	0.72	0.05	0.49	0.34
3 months	8.2	(5.2)	10.2	(7.0)	7.9	(6.3)	14.0	(11.6)							
3 years	5.3	(5.4)	9.8	(5.2)	7.3	(5.9)	12.7	(15.0)							
SCL-90 Global Severity Index															
Before	5.7	(3.8)	11.3	(4.0)	4.9	(2.2)	16.6	(9.8)	1.27	6.73*	2.49	0.40	0.08	2.47†	0.24
3 months	7.8	(3.6)	8.0	(6.9)	6.6	(4.0)	15.0	(12.5)							
3 years	6.5	(5.0)	8.0	(6.6)	5.2	(5.5)	15.3	(13.6)							
SCL-90 Positive Symptom Total															
Before	5.3	(3.7)	11.7	(3.5)	4.7	(2.0)	12.4	(9.7)	0.15	4.28*	0.07	0.28	0.36	0.79	0.09
3 months	6.3	(2.8)	9.5	(6.8)	6.8	(3.1)	12.4	(10.7)							
3 years	4.7	(4.6)	9.8	(5.7)	5.8	(7.4)	11.5	(11.8)							
Concept Level Analogy Test															
Before	13.0	(3.4)	6.0	(4.2)	10.1	(12.3)	8.0	(4.5)	0.20	2.12	0.82	0.09	1.00	0.90	0.08
3 months	11.0	(6.7)	4.5	(2.3)	10.1	(12.9)	10.4	(6.4)							
3 years	12.3	(3.8)	3.2	(4.1)	11.6	(15.1)	8.3	(6.2)							
Jenkins Type A Overall Score															
Before	9.8	(4.2)	7.9	(5.2)	9.9	(11.0)	9.4	(6.6)	0.35	0.84	0.03	0.53	0.29	1.99	0.42

3 months	10.0	(5.8)	7.3	(5.4)	10.6	(11.9)	10.2	(6.3)							
3 years	9.5	(5.5)	5.3	(4.8)	12.4	(13.7)	7.0	(6.6)							
Jenkins Type A Speed Impatience Subscale															
Before	10.2	(4.3)	7.6	(4.9)	10.8	(10.3)	8.0	(6.1)	0.24	0.89	0.00	0.60	0.28	0.26	0.08
3 months	10.7	(7.1)	6.7	(3.1)	11.5	(11.3)	8.9	(4.6)							
3 years	8.2	(5.9)	6.4	(5.3)	10.3	(13.1)	8.4	(6.3)							
Jenkins Type A Job Involvement Subscale															
Before	7.4	(4.4)	9.6	(4.9)	5.8	(1.7)	12.7	(10.2)	0.39	2.00	0.17	0.35	0.33	0.34	0.61
3 months	5.8	(3.1)	10.3	(6.1)	8.0	(2.2)	12.3	(11.5)							
3 years	6.2	(5.0)	7.9	(5.8)	7.3	(6.4)	11.5	(12.8)							
Jenkins Type A Hard Driving Competitiveness Subscale															
Before	9.3	(3.2)	8.3	(5.9)	12.7	(10.8)	6.3	(3.7)	0.30	2.24	0.70	0.73	0.25	3.09*	0.51
3 months	8.8	(6.0)	8.3	(5.5)	11.7	(11.0)	9.0	(6.7)							
3 years	9.2	(5.3)	5.6	(5.3)	13.9	(12.5)	4.6	(5.1)							
Recent Life Changes Number Endorsed															
Before	7.0	(4.5)	8.4	(5.5)	6.6	(5.3)	12.7	(11.3)	0.66	0.48	2.33	1.07	0.15	2.03	2.25
3 months	11.6	(5.4)	5.2	(4.2)	8.4	(6.9)	14.6	(11.1)							
3 years	8.8	(6.3)	6.0	(5.7)	5.8	(5.3)	14.1	(14.4)							
Recent Life Changes Scaled Total															
Before	6.8	(5.1)	8.4	(5.5)	6.3	(5.3)	13.8	(10.3)	0.71	0.54	1.98	0.42	0.03	2.37	0.77
3 months	11.3	(6.1)	5.2	(4.2)	8.1	(6.9)	13.5	(10.8)							
3 years	8.0	(7.0)	6.0	(5.7)	6.3	(4.8)	13.8	(13.9)							

(continued)

Table A-1. *(Continued)*

Intake measure	Treatment				Control				F-Ratio[a]						
	Low IM		High IM		Low IM		High IM								
	$\bar{X}$	*SD*	$\bar{X}$	*SD*	$\bar{X}$	*SD*	$\bar{X}$	*SD*	G	IM	GIM	T	TG	TIM	TGIM
Recent Life Changes Subjective Totals															
Before	6.3	(4.6)	8.0	(0)	7.3	(5.8)	8.3	(5.0)	0.92	0.11	0.29	0.13	0.62	0.90	0.52
3 months	7.8	(6.2)	1.0	(0)	9.2	(7.6)	9.0	(3.2)							
3 years	6.3	(6.9)	4.0	(0)	7.7	(4.8)	8.7	(7.0)							
Left Main Coronary Disease															
Before	8.8	(5.1)	8.8	(4.7)	9.8	(9.9)	9.7	(7.9)	0.28	0.20	0.00	0.46	0.54	3.12*	0.01
3 months	11.2	(6.4)	6.9	(4.5)	12.1	(11.4)	7.8	(3.9)							
3 years	6.8	(6.6)	7.5	(5.0)	9.9	(12.9)	10.0	(7.2)							
Age															
Before	8.2	(5.5)	10.0	(1.8)	13.4	(10.9)	7.0	(5.9)	0.45	0.50	1.58	0.72	0.38	1.20	0.13
3 months	7.7	(5.8)	10.5	(4.8)	12.3	(11.9)	9.1	(6.8)							
3 years	7.2	(6.1)	7.3	(4.1)	13.6	(14.9)	6.7	(5.5)							
Prior Infarction															
Before	9.0	(5.1)	8.6	(4.6)	11.0	(11.5)	8.8	(6.9)	0.52	0.41	0.18	0.42	0.35	0.10	0.03
3 months	9.0	(5.4)	8.1	(6.0)	12.6	(12.2)	8.9	(6.7)							
3 years	7.5	(5.4)	7.0	(5.8)	11.7	(15.6)	8.6	(5.6)							
History of Hypertension															
Before	11.5	(4.9)	7.6	(4.2)	6.0	(3.8)	12.7	(10.8)	0.15	0.00	1.05	0.17	0.64	0.67	3.44*
3 months	8.5	(3.5)	8.6	(6.4)	10.4	(7.0)	10.6	(11.2)							
3 years	10.5	(2.5)	5.8	(5.8)	8.4	(7.1)	11.1	(13.3)							

Family History of Heart Disease															
Before	8.9	(5.1)	8.7	(4.5)	9.7	(9.9)	9.8	(7.5)	0.37	0.12	0.12	0.82	0.23	2.19	0.51
3 months	6.0	(4.1)	11.5	(5.6)	10.0	(10.2)	11.6	(8.1)							
3 years	6.9	(5.0)	7.7	(6.3)	10.5	(12.4)	8.8	(7.1)							
Triple Coronary Disease															
Before	9.0	(5.3)	8.5	(4.3)	10.2	(10.3)	8.8	(5.5)	0.19	0.28	0.31	0.46	0.13	0.48	0.97
3 months	9.3	(4.4)	7.7	(6.9)	11.7	(10.8)	7.8	(4.8)							
3 years	6.1	(5.3)	8.5	(5.7)	11.3	(12.5)	7.0	(5.4)							
Number of Months with Angina															
Before	9.6	(5.3)	7.4	(3.5)	5.0	(2.1)	16.0	(10.6)	1.3	1.7	5.2*	0.30	0.35	0.55	0.19
3 months	9.8	(6.5)	6.6	(2.9)	7.5	(2.6)	14.9	(13.0)							
3 years	8.1	(6.1)	5.8	(4.2)	6.1	(6.1)	15.4	(13.5)							
Number of Bypasses															
Before	8.1	(3.7)	9.8	(6.2)	8.2	(6.5)	14.5	(14.2)	3.37†	5.59*	4.78*	0.60	2.66†	1.74	5.74**
3 months	7.9	(6.3)	9.6	(4.3)	7.3	(4.0)	20.3	(14.4)							
3 years	8.0	(5.7)	6.0	(5.1)	5.8	(4.9)	22.3	(15.0)							
Systolic Blood Pressure															
Before	9.7	(5.5)	8.0	(4.1)	6.7	(5.1)	12.1	(10.7)	0.35	1.04	0.28	0.31	0.56	2.02	1.75
3 months	6.3	(4.4)	10.4	(6.0)	7.0	(4.1)	13.2	(11.5)							
3 years	6.3	(4.9)	8.0	(6.1)	9.1	(6.1)	10.8	(13.8)							
Diastolic Blood Pressure															
Before	7.8	(6.3)	9.7	(3.3)	10.3	(11.4)	9.3	(7.2)	0.56	0.27	0.85	0.70	0.59	1.50	0.57
3 months	8.9	(5.6)	8.8	(6.0)	14.1	(13.2)	7.7	(3.5)							
3 years	6.2	(5.5)	7.7	(5.9)	12.9	(14.5)	7.7	(6.9)							

(continued)

Table A-1. (*Continued*)

	Treatment				Control				*F*-Ratio[a]						
	Low IM		High IM		Low IM		High IM								
Intake measure	$\bar{X}$	*SD*	$\bar{X}$	*SD*	$\bar{X}$	*SD*	$\bar{X}$	*SD*	G	IM	GIM	T	TG	TIM	TGIM
Ejection Fraction															
Before	10.2	(4.4)	7.7	(5.3)	12.3	(10.6)	8.8	(7.1)	0.66	0.84	0.00	0.76	0.25	0.52	0.11
3 months	10.0	(7.6)	7.9	(4.3)	12.1	(11.2)	11.2	(7.2)							
3 years	9.4	(6.7)	5.1	(4.1)	12.3	(14.4)	8.9	(6.0)							
New York Heart Association Functional Class															
Before	5.6	(4.1)	11.8	(3.9)	7.0	(6.0)	12.6	(11.2)	0.69	2.93	0.00	0.49	0.71	0.96	0.03
3 months	6.3	(2.8)	9.4	(6.2)	8.7	(7.9)	12.4	(11.6)							
3 years	3.2	(3.4)	9.4	(3.8)	6.8	(5.4)	13.5	(14.1)							
PRE	III		IV		III		IV								
MBHI Introversive															
Before	12.3	(4.5)	5.5	(5.3)	10.0	(10.5)	5.2	(1.9)	0.00	2.74	0.02	0.21	0.36	0.35	0.19
3 months	11.0	(9.5)	5.8	(1.9)	11.1	(11.0)	5.0	(3.8)							
3 years	11.3	(6.7)	2.8	(3.6)	11.3	(13.7)	4.6	(3.4)							
MBHI Inhibited															
Before	4.8	(4.4)	13.3	(3.2)	5.4	(2.3)	12.2	(12.5)	0.09	5.03*	0.00	0.14	0.83	0.70	1.26
3 months	4.5	(3.0)	12.7	(7.0)	6.6	(4.3)	12.0	(13.6)							
3 years	3.3	(3.6)	10.7	(7.8)	3.9	(2.6)	15.7	(15.3)							
MBHI Cooperative															
Before	8.5	(0.7)	8.4	(7.2)	9.2	(12.8)	7.6	(4.6)	0.19	0.08	0.56	0.10	0.89	1.74	1.46
3 months	4.0	(4.2)	9.6	(6.7)	11.0	(14.1)	7.4	(3.9)							
3 years	6.0	(2.8)	6.6	(7.8)	13.5	(16.3)	5.5	(4.6)							

MBHI Sociable															
Before	12.7	(4.0)	5.3	(5.1)	13.0	(11.8)	4.8	(2.4)	0.13	8.55**	0.03	0.10	0.63	1.60	0.10
3 months	13.0	(6.6)	4.3	(2.8)	14.0	(12.6)	5.1	(3.4)							
3 years	12.7	(4.5)	1.8	(1.7)	16.5	(14.6)	3.3	(2.1)							
MBHI Confident															
Before	9.6	(6.3)	5.5	(4.9)	13.0	(11.8)	4.8	(2.4)	0.40	2.38	0.37	0.11	0.36	0.02	0.09
3 months	9.2	(7.2)	5.0	(2.8)	13.7	(12.8)	5.4	(3.6)							
3 years	7.2	(7.4)	4.5	(4.9)	14.0	(15.3)	5.1	(6.0)							
MBHI Forceful															
Before	8.4	(7.2)	8.5	(0.7)	7.5	(5.5)	8.9	(10.8)	0.09	0.02	0.38	0.22	0.52	0.54	1.19
3 months	9.6	(6.7)	4.0	(4.2)	6.7	(4.4)	10.6	(12.0)							
3 years	6.6	(7.8)	6.0	(2.8)	6.3	(4.8)	10.9	(14.7)							
MBHI Respectful															
Before	6.6	(5.4)	13.0	(5.7)	13.8	(14.3)	6.0	(4.6)	0.18	0.14	2.30	0.20	0.51	0.74	0.12
3 months	7.4	(7.4)	9.5	(3.5)	15.5	(15.9)	6.3	(4.1)							
3 years	4.8	(7.0)	10.5	(3.5)	13.8	(20.2)	7.0	(6.1)							
MBHI Sensitive															
Before	5.5	(5.3)	12.3	(4.5)	5.1	(2.2)	12.5	(12.3)	0.11	4.35†	0.03	0.11	0.57	0.94	0.03
3 months	5.8	(1.9)	11.0	(9.5)	6.0	(3.4)	12.8	(13.5)							
3 years	2.8	(3.6)	11.3	(6.7)	4.5	(2.7)	14.8	(16.0)							
MBHI Chronic Tension															
Before	5.3	(5.1)	12.7	(4.0)	5.6	(2.3)	9.8	(10.6)	0.01	3.48†	0.26	0.31	0.28	0.96	0.05
3 months	4.3	(2.8)	13.0	(6.6)	6.4	(3.4)	10.3	(11.5)							
3 years	1.8	(1.7)	12.7	(4.5)	4.2	(1.8)	11.6	(13.7)							
MBHI Recent Stress															
Before	6.0	(4.7)	14.5	(3.5)	5.2	(1.7)	10.6	(11.1)	0.19	5.13*	0.24	0.16	0.20	0.74	0.07
3 months	4.8	(2.7)	16.0	(5.7)	5.3	(3.5)	11.6	(11.7)							
3 years	3.0	(3.2)	15.0	(2.8)	4.0	(3.6)	12.6	(14.0)							
MBHI Pessimism Premorbid															
Before	6.0	(4.7)	14.5	(3.5)	5.1	(2.4)	16.3	(13.7)	0.26	13.48**	0.16	0.05	0.93	1.99	0.70
3 months	4.8	(2.7)	16.0	(5.7)	5.9	(3.4)	16.5	(15.5)							
3 years	3.0	(3.2)	15.0	(2.8)	3.9	(2.8)	21.5	(15.7)							

(continued)

Table A-1. *(Continued)*

	Treatment				Control				*F*-Ratio[a]						
	Low IM		High IM		Low IM		High IM								
Intake measure	$\bar{X}$	*SD*	$\bar{X}$	*SD*	$\bar{X}$	*SD*	$\bar{X}$	*SD*	G	IM	GIM	T	TG	TIM	TGIM
MBHI Future Dispair															
Before	4.3	(5.9)	11.5	(4.0)	5.3	(2.0)	15.8	(14.4)	0.91	4.72*	0.07	0.19	0.75	1.52	0.12
3 months	5.3	(2.1)	10.0	(8.0)	7.1	(4.0)	13.5	(17.3)							
3 years	1.0	(1.0)	10.5	(5.7)	6.0	(5.5)	16.3	(19.3)							
MBHI Social Alienation															
Before	4.3	(5.9)	11.5	(4.0)	4.9	(2.5)	12.8	(11.9)	0.37	4.50*	0.02	0.22	0.89	2.73†	0.06
3 months	5.3	(2.1)	10.0	(8.0)	6.8	(4.2)	11.8	(13.7)							
3 years	1.0	(1.0)	10.5	(5.7)	4.0	(2.6)	15.5	(15.4)							
MBHI Somatic Anxiety															
Before	5.3	(5.1)	12.7	(4.0)	5.1	(2.2)	12.5	(12.3)	0.09	6.02*	0.01	0.11	0.53	0.95	0.07
3 months	4.3	(2.8)	13.0	(6.6)	6.0	(3.4)	12.8	(13.5)							
3 years	1.8	(1.7)	12.7	(4.5)	4.5	(2.7)	14.8	(16.0)							
MBHI Allergic Inclination															
Before	5.3	(5.1)	12.7	(4.0)	5.6	(1.9)	11.0	(11.9)	0.01	3.97†	0.20	0.16	0.43	0.29	0.18
3 months	4.3	(2.8)	13.0	(6.6)	6.0	(3.7)	11.9	(12.6)							
3 years	1.8	(1.7)	12.7	(4.5)	6.0	(6.0)	11.9	(15.1)							
MBHI Gastrointestinal Susceptibility															
Before	5.3	(5.1)	12.7	(4.0)	5.1	(2.2)	12.5	(12.3)	0.09	6.02*	0.01	0.11	0.53	0.95	0.07
3 months	4.3	(2.8)	13.0	(6.6)	6.0	(3.4)	12.8	(13.5)							
3 years	1.8	(1.7)	12.7	(4.5)	4.5	(2.7)	14.8	(16.0)							

MBHI Cardiovascular Tendency															
Before	5.3	(5.1)	12.7	(4.0)	5.2	(1.9)	10.0	(10.5)	0.01	4.22†	01.3	0.24	0.29	0.74	0.02
3 months	4.3	(2.8)	13.0	(6.6)	4.8	(3.7)	11.2	(11.0)							
3 years	1.8	(1.7)	12.7	(4.5)	4.0	(2.0)	11.7	(13.6)							
MBHI Pain Proneness															
Before	7.5	(3.9)	9.7	(8.7)	4.8	(2.3)	14.6	(12.5)	0.31	4.30†	0.24	0.23	0.43	0.60	1.11
3 months	4.3	(2.8)	13.0	(6.6)	5.8	(3.6)	14.6	(14.1)							
3 years	3.8	(3.1)	10.0	(8.9)	5.7	(5.8)	14.8	(17.0)							
MBHI Life Threat Reactivity															
Before	5.3	(5.1)	12.7	(4.0)	5.4	(2.2)	13.4	(13.5)	0.29	7.52*	0.01	0.02	0.94	2.31	0.40
3 months	4.3	(2.8)	13.0	(6.6)	6.3	(3.3)	13.6	(14.9)							
3 years	1.8	(1.7)	12.7	(4.5)	4.0	(3.0)	17.8	(15.9)							
MBHI Emotional Vulnerability															
Before	3.7	(4.7)	12.0	(3.7)	4.2	(1.9)	10.6	(10.2)	0.05	1.89	0.03	0.29	0.54	1.08	0.02
3 months	5.7	(2.3)	9.8	(8.2)	7.0	(4.7)	10.0	(11.4)							
3 years	3.0	(4.4)	9.0	(7.2)	5.8	(3.3)	10.7	(14.0)							

[a]G = Group (treatment versus control), T = time, IM = Intake Measure.
†$p < .10$. *$p < .05$. **$p < .01$.

interaction with group was number of months with angina. In the treatment group those who had had angina many months were less depressed, whereas in the control group those who had angina many months were more depressed. This effect occurred across all three times of evaluation.

A marginally significant trend for GSI by time suggests that patients in both treatment and control groups who had a low GSI score before surgery became slightly more depressed after surgery and those with high GSI scores before surgery became less depressed after surgery. There was also a significant trend for the Jenkins Hard Driving–Competitiveness Subscale by time. Type A (shown in the low IM column) showed little change in depression across the three times of evaluation, although there was a slight dip at 3 months. Type B (shown in the high IM column) had a slight rise in depression at 3 months, then a reduction at 3 years. There was a significant time by Left Main Coronary Disease interaction. For those who had left main coronary disease (shown in the low IM column) there was a peak in depression at 3 months, whereas for those not having the disease (shown in the high IM column) there was a dip in depression at 3 months.

The significant triple interaction between History of Hypertension, Group, and time is pictured in Figure A-1. Treatment patients with a history of hypertension became less depressed at 3 months, then more depressed again at 3 years, whereas treatment patients without a history of hypertension did just the opposite, becoming slightly more depressed at 3 months but less so at 3 years. Control patients with a history of hypertension showed a sizeable rise in depression at 3 months, then a drop at 3 years. Control patients without a history of hypertension became less depressed at 3 months but slightly more depressed at 3 years. Figure A-2 displays the significant triple interaction for Number of Bypasses by Group by time. Those control patients with many bypasses began at a clinically high level of depression and became more depressed at 3 months and 3 years. Treatment patients with many bypasses stayed about the same at 3 months but became better at 3 years. Control patients with few bypasses showed a small but steady improvement across time.

Table A-2 displays the results of the 2 × 2 × 2 mixed-design ANOVA (Group by Intake Measure by time—before surgery and 3 months after surgery) with Beck Depression Inventory as the dependent variable. Patients with higher levels on GSI, PST, and LOC externality and a greater number of bypasses all had significantly higher levels of depression both before and after surgery in both treatment and control groups. Additionally, patients with New York Heart Association Functional Class IV had significantly more depression at both times than those in Class II or III. Similarly patients with a Type B score (i.e., the High IM columns on Table A-2) on the Jenkins Job Involvement Subscale and patients who endorsed many recent life changes had a marginally significant tendency to have higher levels of depression. As in Table A-1, many of the MBHI scales had main effects in Table A-2. For Introversive Personality Style, Sociable Personality Style, and Confident Personality Style a high score presurgically tended to be associated with a low BDI score at all three assessments in both treatment and

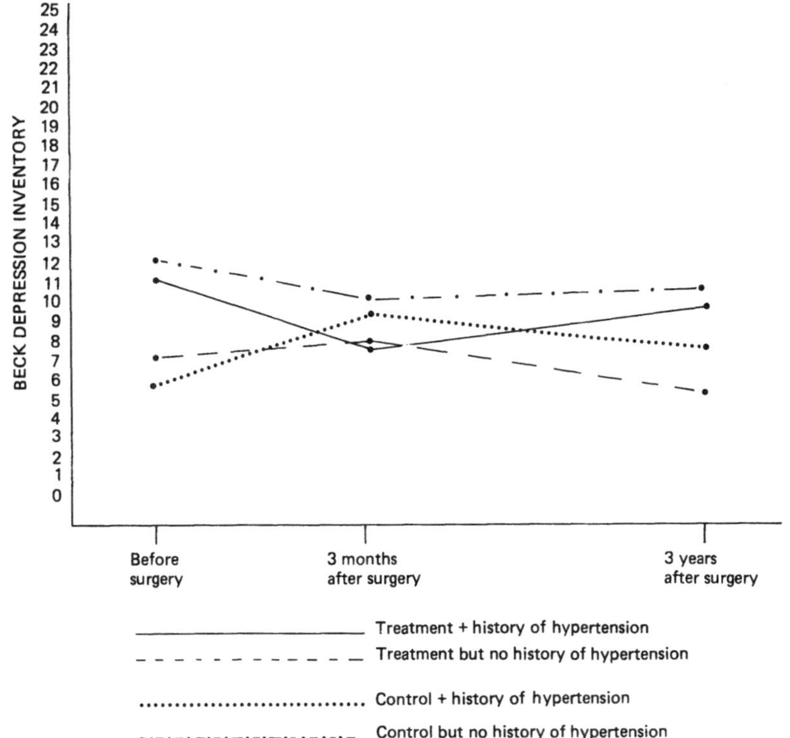

Figure A-1. Beck Depression Inventory by time, Group, and History of Hypertension.

control groups. On the other hand, low presurgical scores on Inhibited Personality Style, Sensitive Personality Style, Recent Stress, Premorbid Pessimism, Future Despair, Social Alienation, Somatic Anxiety, Allergic Inclination, Gastrointestinal Susceptibility, Pain Proneness, Life Threat Reactivity, Emotional Vulnerability, and, to a marginal degree, Cardiovascular Tendency all tended to be associated with low BDI scores.

Three intake measures interacted with Group significantly and one more to a marginally significant extent. Older patients were more depressed than younger patients if they were in the treatment group but less depressed if they were in the control group. If they had a history of hypertension, patients were more depressed in the treatment group and less depressed in the control group. While a greater number of months suffering from angina tended to result in marginally higher levels of depression in the control group, this was not true in the treatment group. The Number of Recent Life Changes Endorsed interacted with time as well as with Group. The result of these two two-way interactions is very similar to that of a single triple interaction. Figure A-3 displays these interactions. Fewer changes endorsed before surgery tended to result in greater

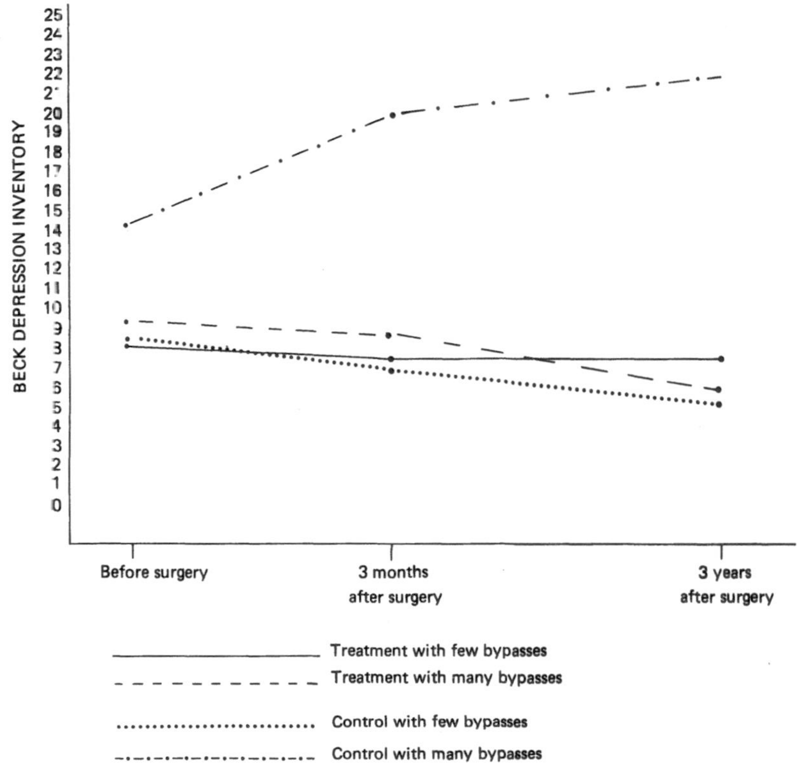

Figure A-2. Beck Depression Inventory by time, Group, and Number of Bypasses.

increases in depression from before to after surgery in both treatment and control groups. If many changes were endorsed before surgery, however, treatment subjects became less depressed, whereas control subjects remained at the same clinically high level of depression.

GSI also had a significant interaction with time. Those with low GSI scores before surgery became more depressed from before to after surgery and those with higher levels of GSI became less depressed. Similarly the subjective total score on Recent Life Changes had a marginal interaction with time. Those reporting less presurgical stress became more depressed whereas those reporting more stress became less depressed. On the MBHI the Introversive Personality Style had a significant interaction with time such that patients in the low introversive group tended to improve and those in the high introversive group tended to become slightly depressed. The Sensitive Personality had a marginal significance such that patients with higher Sensitive scales tended to become less depressed and those with lower Sensitive scores tended to become more depressed.

Table A-2. Means, Standard Deviations, and *F*-Ratios of Group by Various Intake Measures for the Beck Depression Inventory before and 3 Months after Surgery

	Treatment				Control				*F*-Ratio[a]						
	Low IM		High IM		Low IM		High IM								
Intake Measure	$\bar{X}$	*SD*	$\bar{X}$	*SD*	$\bar{X}$	*SD*	$\bar{X}$	*SD*	G	IM	GIM	T	TG	TIM	TGIM
Locus of Control															
Before	6.9	(4.7)	9.7	(6.3)	6.7	(5.2)	14.5	(10.2)	1.81	7.12**	0.97	0.02	0.00	0.91	0.96
3 months	6.7	(4.1)	9.6	(7.0)	8.2	(6.0)	12.8	(8.6)							
SCL-90 Global Severity Index															
Before	4.9	(3.4)	12.2	(4.8)	5.7	(3.4)	15.5	(9.9)	3.53†	24.04***	1.92	0.47	0.47	4.63*	0.78
3 months	6.2	(2.9)	8.7	(6.8)	6.7	(3.6)	14.5	(9.1)							
SCL-90 Positive Symptom Total															
Before	4.5	(3.3)	12.0	(4.6)	6.0	(3.8)	13.3	(9.8)	1.65	16.26***	0.11	0.11	0.53	1.87	0.51
3 months	5.4	(2.3)	9.4	(6.5)	6.9	(2.6)	13.1	(8.7)							
Concept Level Analogy Test															
Before	11.1	(5.9)	8.0	(5.8)	10.8	(11.2)	10.1	(6.4)	1.25	0.98	0.54	1.05	2.26	0.02	0.10
3 months	9.1	(6.1)	5.1	(2.7)	11.1	(9.7)	10.8	(6.4)							
Jenkins Type A Overall Score															
Before	9.6	(4.0)	6.6	(5.0)	11.4	(9.6)	10.5	(9.4)	3.20†	0.69	0.09	0.37	0.37	0.16	0.35
3 months	7.8	(5.2)	6.5	(4.5)	11.5	(9.6)	10.4	(6.0)							
Jenkins Type A Speed Impatience Subscale															
Before	9.9	(3.8)	7.7	(7.2)	11.8	(9.2)	9.8	(9.8)	1.65	1.51	0.00	0.32	0.27	0.08	0.00
3 months	9.2	(7.1)	6.5	(3.3)	12.0	(9.5)	9.5	(4.9)							

(*continued*)

Table A-2. *(Continued)*

	Treatment				Control				*F*-Ratio[a]						
	Low IM		High IM		Low IM		High IM								
Intake Measure	$\bar{X}$	*SD*	$\bar{X}$	*SD*	$\bar{X}$	*SD*	$\bar{X}$	*SD*	G	IM	GIM	T	TG	TIM	TGIM
Jenkins Type A Job Involvement Subscale															
Before	7.9	(3.6)	8.3	(5.3)	6.1	(1.9)	12.4	(9.2)	1.81	2.94[†]	0.49	0.05	2.67	0.09	4.26*
3 months	5.0	(2.6)	8.3	(5.4)	9.1	(3.1)	11.7	(9.5)							
Jenkins Type A Hard Driving Competitiveness Subscale															
Before	8.4	(2.7)	7.9	(6.1)	13.9	(11.1)	8.0	(6.2)	3.26[†]	1.31	1.46	0.43	0.43	1.94	0.35
3 months	6.7	(4.9)	7.5	(4.9)	12.3	(9.4)	9.6	(6.1)							
Recent Life Changes Number Endorsed															
Before	7.2	(5.6)	8.9	(6.4)	6.5	(4.8)	14.8	(10.8)	3.76[†]	2.88[†]	5.01*	0.01	1.47	5.11*	1.09
3 months	9.0	(5.3)	5.3	(3.6)	8.6	(6.0)	14.9	(8.5)							
Recent Life Changes Scaled Total															
Before	7.1	(5.9)	8.9	(6.4)	8.2	(7.4)	12.0	(8.4)	2.78	0.60	1.36	0.00	1.79	2.58	2.12
3 months	8.6	(5.4)	5.3	(3.6)	9.4	(6.7)	13.0	(8.8)							
Recent Life Changes Subjective Totals															
Before	6.6	(5.6)	10.3	(6.3)	8.9	(8.3)	9.2	(4.3)	1.23	0.03	0.15	1.05	2.10	3.40[†]	1.12
3 months	7.0	(5.5)	5.3	(3.6)	10.0	(7.7)	8.9	(2.7)							

Left Main Coronary Disease															
Before	8.4	(4.1)	8.4	(6.0)	11.6	(11.6)	9.6	(5.9)	1.47	1.64	0.01	0.15	0.00	2.16	0.86
3 months	10.9	(7.6)	6.6	(3.8)	12.4	(10.5)	9.4	(4.8)							
Age															
Before	7.9	(6.0)	8.8	(4.5)	13.6	(11.0)	8.2	(5.6)	2.86	0.52	4.04*	0.01	0.08	1.74	0.04
3 months	6.4	(4.5)	9.7	(6.0)	12.9	(9.3)	9.3	(5.8)							
Prior Infarction															
Before	10.1	(6.2)	6.9	(4.4)	10.0	(9.9)	10.8	(8.5)	2.07	0.34	0.13	0.00	0.53	0.02	2.40
3 months	7.9	(4.7)	7.7	(6.1)	11.8	(10.8)	11.2	(5.8)							
History of Hypertension															
Before	12.4	(5.8)	6.8	(4.5)	7.8	(4.0)	12.8	(11.2)	0.72	0.02	5.36*	0.15	0.69	0.00	1.82
3 months	10.0	(6.2)	7.0	(4.9)	9.4	(5.5)	12.0	(9.6)							
Family History of Heart Disease															
Before	7.4	(4.4)	7.7	(4.2)	11.0	(9.9)	9.0	(6.9)	1.92	0.01	0.71	0.40	0.03	1.09	0.23
3 months	6.4	(5.1)	9.6	(5.6)	11.2	(8.6)	10.3	(7.1)							
Triple Coronary Disease															
Before	8.8	(6.6)	7.9	(3.7)	9.6	(8.4)	12.5	(10.1)	2.12	0.03	0.02	0.33	0.01	0.65	3.18†
3 months	7.4	(4.3)	8.3	(6.7)	11.3	(9.0)	9.6	(5.3)							
Number of Months with Angina															
Before	8.1	(4.6)	8.9	(6.7)	6.9	(4.0)	13.8	(10.6)	2.24	2.40	3.52†	0.06	0.55	1.97	0.03
3 months	8.6	(6.2)	6.6	(3.7)	8.5	(2.9)	13.1	(10.0)							
Number of Bypasses															
Before	8.1	(5.2)	9.1	(6.4)	9.4	(5.9)	12.7	(13.0)	3.36†	4.04*	1.97	0.11	0.89	2.38	2.22
3 months	7.5	(5.8)	8.6	(4.5)	7.8	(3.8)	16.6	(10.5)							

(continued)

Table A-2. (*Continued*)

	Treatment				Control				*F*-Ratio[a]						
	Low IM		High IM		Low IM		High IM								
Intake Measure	$\bar{X}$	*SD*	$\bar{X}$	*SD*	$\bar{X}$	*SD*	$\bar{X}$	*SD*	G	IM	GIM	T	TG	TIM	TGIM
Systolic Blood Pressure															
Before	9.2	(6.2)	7.6	(4.6)	7.5	(4.9)	12.7	(10.5)	1.65	1.79	2.82	0.02	0.27	0.24	0.53
3 months	7.6	(5.7)	8.1	(5.3)	8.0	(5.2)	12.8	(9.1)							
Diastolic Blood Pressure															
Before	8.9	(6.5)	7.9	(4.7)	10.2	(9.0)	10.8	(9.1)	2.09	0.56	0.00	0.00	0.29	1.61	1.15
3 months	8.6	(6.3)	7.2	(4.7)	12.7	(10.6)	9.2	(4.4)							
Ejection Fraction															
Before	10.5	(6.0)	6.8	(4.5)	12.1	(9.5)	11.0	(9.8)	2.67	1.06	0.00	0.03	0.07	0.18	1.88
3 months	8.3	(5.7)	8.1	(5.7)	13.1	(10.2)	10.2	(6.4)							
New York Heart Association Functional Class	II		III		II		III								
Before	6.5	(3.7)	5.9	(3.6)	10.5	(3.5)	7.5	(5.0)	1.77	4.54*	0.01	0.67	0.01	0.92	0.34
3 months	5.8	(2.2)	5.7	(2.4)	7.5	(3.5)	9.3	(6.7)							
	IV						IV								
Before	11.7	(6.7)					15.1	(12.1)							
3 months	10.2	(6.8)					13.2	(10.4)							
MBHI Introversive															
Before	14.4	(6.3)	6.4	(5.7)	12.8	(10.5)	6.0	(3.3)	0.00	4.81*	0.00	0.79	0.85	3.81†	0.49
3 months	9.6	(7.7)	7.3	(6.0)	11.5	(8.9)	7.4	(6.0)							
MBHI Inhibited															
Before	6.3	(5.6)	12.1	(7.3)	6.5	(3.6)	14.6	(11.5)	0.48	8.53**	0.00	0.36	0.04	0.33	0.91
3 months	5.1	(3.6)	11.7	(7.5)	7.7	(5.3)	12.6	(9.9)							
MBHI Cooperative															
Before	9.8	(7.9)	8.3	(6.5)	10.5	(9.8)	10.6	(9.1)	0.70	0.01	0.31	1.18	0.32	0.86	4.88*
3 months	5.0	(3.9)	9.8	(7.0)	11.2	(10.2)	8.8	(4.8)							

MBHI Sociable															
Before	11.3	(7.6)	7.0	(6.0)	16.0	(12.7)	7.3	(4.3)	1.41	9.51**	0.39	0.40	0.00	0.02	0.69
3 months after	11.6	(7.7)	5.2	(3.5)	14.7	(10.5)	7.4	(5.0)							
MBHI Confident															
Before	9.7	(6.0)	7.9	(8.2)	14.8	(11.1)	6.3	(4.3)	0.92	5.80*	0.63	0.54	0.11	0.04	2.26
3 months after	10.3	(7.3)	5.0	(3.4)	13.0	(9.5)	7.3	(5.5)							
MBHI Forceful															
Before	8.3	(6.5)	9.8	(7.9)	12.2	(10.0)	9.2	(8.8)	0.80	0.57	0.01	1.13	0.33	1.18	4.12*
3 months	9.8	(7.0)	5.0	(3.9)	10.7	(6.6)	9.6	(9.5)							
MBHI Respectful															
Before	7.7	(5.6)	11.4	(9.3)	15.4	(11.5)	7.6	(6.4)	0.97	0.93	2.58	1.35	0.16	0.06	3.04†
3 months	8.2	(7.3)	7.6	(4.3)	13.0	(10.8)	8.4	(5.8)							
MBHI Sensitive															
Before	6.7	(5.9)	13.6	(7.0)	5.7	(3.1)	14.6	(10.9)	0.10	6.41*	0.22	0.84	0.46	3.82†	0.00
3 months	7.2	(6.0)	9.8	(7.5)	7.5	(5.1)	12.4	(9.6)							
MBHI Chronic Tension															
Before	6.6	(5.8)	13.8	(6.8)	9.6	(10.1)	11.1	(9.1)	0.05	2.60	0.51	0.80	0.14	0.12	0.85
3 months after	6.6	(6.3)	11.0	(6.1)	8.3	(4.9)	11.2	(9.6)							
MBHI Recent Stress															
Before	6.5	(5.5)	16.0	(5.9)	7.8	(7.1)	12.2	(10.2)	0.04	5.48*	0.32	1.12	0.47	0.94	0.98
3 months after	6.8	(6.0)	11.8	(6.9)	7.3	(5.6)	11.8	(9.1)							
MBHI Premorbid Pessimism															
Before	5.9	(4.8)	15.4	(6.3)	5.4	(3.1)	15.7	(10.6)	0.05	17.26***	0.11	0.79	0.25	1.64	0.08
3 months after	6.1	(5.9)	12.2	(5.7)	6.1	(3.2)	14.2	(9.6)							
MBHI Future Despair															
Before	5.4	(5.3)	11.6	(7.0)	6.4	(3.4)	16.4	(11.8)	2.08	12.39**	0.40	0.52	0.01	0.57	0.38
3 months after	4.7	(1.7)	10.6	(7.6)	7.1	(3.5)	14.3	(10.9)							
MBHI Social Alienation															
Before	6.9	(6.4)	10.9	(7.1)	6.4	(3.7)	15.5	(11.5)	0.99	6.47*	0.37	0.52	0.02	0.85	1.37
3 months after	5.8	(3.3)	10.3	(8.0)	7.9	(5.0)	12.7	(10.4)							
MBHI Somatic Anxiety															
Before	5.7	(4.7)	15.8	(5.7)	5.5	(3.1)	14.1	(10.6)	0.03	14.30***	0.04	0.61	0.30	1.31	0.07
3 months after	5.8	(6.0)	12.8	(4.7)	6.2	(3.2)	12.9	(9.5)							

(continued)

Table A-2. *(Continued)*

Intake Measure	Treatment				Control				F-Ratio[a]						
	Low IM		High IM		Low IM		High IM								
	$\bar{X}$	*SD*	$\bar{X}$	*SD*	$\bar{X}$	*SD*	$\bar{X}$	*SD*	G	IM	GIM	T	TG	TIM	TGIM
MBHI Allergic Inclination															
Before	6.2	(4.8)	12.3	(7.9)	6.8	(3.5)	13.7	(11.5)	0.35	7.15*	0.18	0.46	0.08	0.31	0.26
3 months after	6.3	(6.5)	10.1	(5.9)	6.5	(3.2)	13.2	(9.8)							
MBHI Gastrointestinal Susceptibility															
Before	6.2	(4.8)	12.3	(7.9)	5.5	(3.1)	14.1	(10.6)	0.19	8.51**	0.40	0.37	0.13	0.98	0.01
3 months after	6.3	(6.5)	10.1	(5.9)	6.2	(3.2)	12.9	(9.5)							
MBHI Cardiovascular Tendency															
Before	7.3	(5.7)	11.5	(8.3)	7.0	(3.7)	13.1	(11.3)	0.19	3.59†	0.19	0.41	0.12	0.41	0.01
3 months after	7.0	(6.5)	9.7	(6.3)	7.3	(5.5)	12.1	(9.3)							
MBHI Pain Proneness															
Before	6.6	(4.5)	11.1	(8.3)	5.8	(3.8)	15.3	(10.9)	0.84	9.65**	1.67	0.38	0.05	0.27	0.04
3 months after	6.5	(6.7)	9.5	(5.8)	5.8	(3.1)	14.5	(9.3)							
MBHI Life Threat Reactivity															
Before	5.8	(4.9)	12.0	(7.3)	6.8	(3.8)	15.0	(11.9)	1.20	13.62***	0.10	0.38	0.05	0.08	0.10
3 months after	4.3	(2.1)	11.8	(7.1)	6.4	(2.9)	14.5	(10.1)							
MBHI Emotional Vulnerability															
Before	4.2	(4.6)	11.7	(6.5)	8.2	(7.1)	12.2	(10.5)	1.06	4.49*	0.40	0.19	0.02	0.57	0.07
3 months after	4.7	(2.0)	10.0	(7.3)	8.4	(5.4)	11.4	(9.7)							

[a]G = Group (treatment versus control), T = time, IM = Intake Measure.
†$p < .10$. *$p < .05$. **$p < .01$. ***$p < .001$.

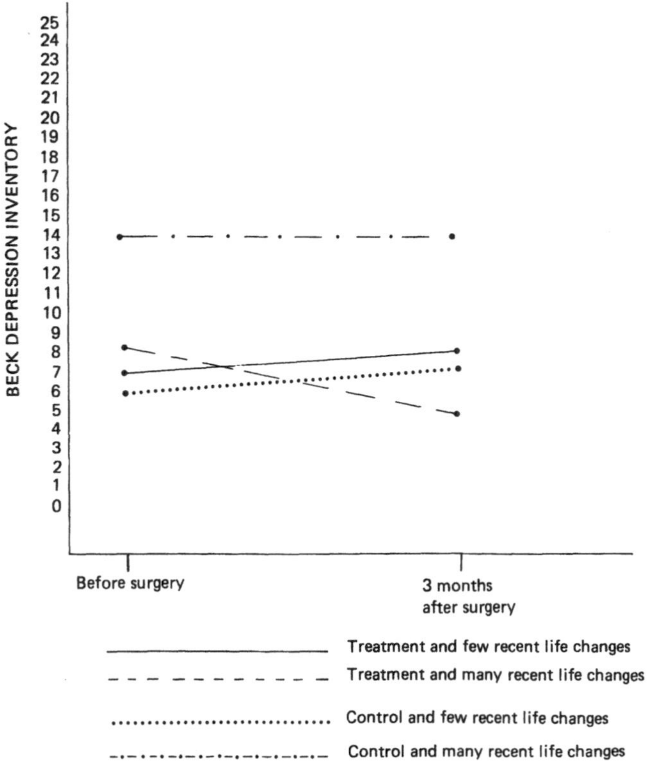

Figure A-3. Beck Depression Inventory by time, Group, and Number of Recent Life Changes Endorsed.

Triple coronary disease had a marginally significant triple interaction. Figure A-4 displays this interaction. Treatment patients with triple coronary disease became slightly less depressed from before to after surgery, whereas control patients with this disease became more depressed. On the other hand, treatment patients without the disease became slightly more depressed and control patients without the disease less depressed. On the MBHI, those patients with low Cooperative or high Forceful Personality Style tended to improve if they were in the crisis intervention counseling treatment whereas other patients showed no consistent change. A similar trend for treatment patients high on the Respectful Personality scale was only marginally significant.

Table A-3 presents the Analogue Depression Scale results at 3 months and 3 years factorially crossed with Group and each of the intake measures. As explained in Chapter 6, this scale was not given presurgically. It also should be remembered that the total number of family member reports of patient depression on this scale is fewer and the variability greater than on the BDI.

History of Hypertension had a significant main effect, with those having

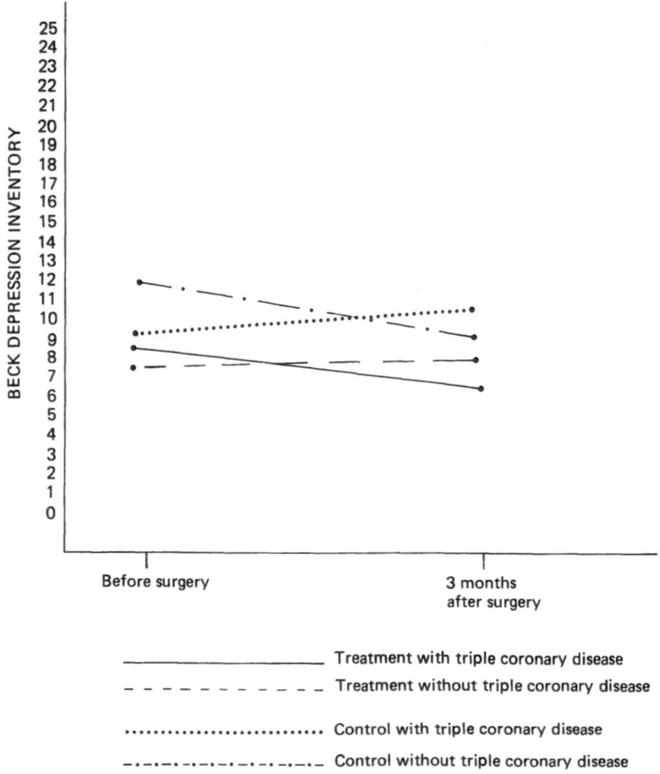

Figure A-4. Beck Depression Inventory by time, Group, and Triple Coronary Disease.

such a history being reported to have more depression than those without such a history. Jenkins Total and Jenkins Job Involvement subscale both had a marginal main effect such that Type A (the column labeled "Low IM" on Table A-3) patients were reported to be more depressed than Type B patients. A low score on the MBHI Inhibited Personality Style, Respectful Personality Style, Chronic Tension, Premorbid Pessimism, Social Alienation, Allergic Inclination, Cardiovascular Tendency, and, to a marginal level, Life Threat Reactivity and Sensitive Personality Style were all associated with a low Analogue score. On the other hand, a high score on the Sociable Personality Style was associated with a low Analogue score.

Locus of Control interacted with Group significantly. Internal treatment patients had lower reported depression than external treatment patients. Internal control patients, on the other hand, had higher reported depression than external control patients. Number of Bypasses had marginally significant two-way interactions with both time and Group. Both treatment groups improved between 3 months and 3 years, whereas both control groups became more de-

Table A-3. Means, Standard Deviations, and F-Ratios of Group by Various Intake Measures for the Analogues before, 3 Months after, and 3 Years after Surgery

	Treatment				Control				F-Ratio[a]						
	Low IM		High IM		Low IM		High IM								
Intake Measure	$\bar{X}$	SD	$\bar{X}$	SD	$\bar{X}$	SD	$\bar{X}$	SD	G	IM	GIM	T	TG	TIM	TGIM
Locus of Control															
Before	N/A	N/A	N/A	N/A	N/A	N/A	N/A	N/A							
3 months	1.5	(1.3)	51.0	(33.8)	6.8	(9.5)	5.6	(6.7)	0.00	2.88	6.15*	0.46	10.86**	2.71	0.52
3 years	0.0	(0.0)	21.5	(25.7)	35.8	(37.6)	23.6	(19.8)							
SCL-90 Global Severity Index															
Before	N/A	N/A	N/A	N/A	N/A	N/A	N/A	N/A							
3 months	24.4	(37.9)	25.8	(30.0)	5.1	(7.5)	8.2	(9.8)	0.00	0.49	0.00	1.30	9.22**	0.54	0.05
3 years	7.0	(15.7)	20.0	(24.8)	17.5	(39.4)	36.8	(14.3)							
SCL-90 Positive Symptom Total															
Before	N/A	N/A	N/A	N/A	N/A	N/A	N/A	N/A							
3 months	30.3	(41.0)	20.8	(28.3)	6.8	(8.1)	5.9	(8.9)	0.00	0.02	0.00	0.95	10.02**	0.41	0.57
3 years	8.8	(17.5)	16.0	(23.2)	32.3	(45.1)	30.0	(17.1)							
Concept Level Analogy Test															
Before	N/A	N/A	N/A	N/A	N/A	N/A	N/A	N/A							
3 months	29.3	(29.8)	5.0	(6.8)	6.7	(8.3)	5.5	(11.0)	0.58	0.72	2.75	3.03	8.47**	0.74	0.26
3 years	20.0	(24.8)	0.0	(0.0)	26.7	(27.8)	42.3	(42.5)							
Jenkins Type A Overall Score															
Before	N/A	N/A	N/A	N/A	N/A	N/A	N/A	N/A							
3 months	47.3	(39.5)	6.5	(6.6)	4.6	(7.7)	8.3	(9.0)	0.03	3.66†	2.63	0.94	12.13**	0.29	2.77
3 years	21.5	(25.7)	4.8	(16.8)	35.0	(38.9)	26.5	(22.9)							

(continued)

Table A-3. *(Continued)*

	Treatment				Control				*F*-Ratio[a]						
	Low IM		High IM		Low IM		High IM								
Intake Measure	$\bar{X}$	*SD*	$\bar{X}$	*SD*	$\bar{X}$	*SD*	$\bar{X}$	*SD*	G	IM	GIM	T	TG	TIM	TGIM
Jenkins Type A Speed Impatience Subscale															
Before	N/A	N/A	N/A	N/A	N/A	N/A	N/A	N/A							
3 months	44.0	(41.4)	8.7	(12.2)	11.3	(8.4)	0.5	(1.2)	0.02	3.77†	1.45	1.13	11.61**	1.56	0.17
3 years	21.5	(25.7)	4.8	(11.8)	31.7	(28.5)	30.3	(37.6)							
Jenkins Type A Job Involvement Subscale															
Before	N/A	N/A	N/A	N/A	N/A	N/A	N/A	N/A							
3 months	34.3	(47.1)	20.7	(26.6)	4.0	(8.4)	8.9	(8.3)	0.01	1.23	0.05	1.46	10.70**	1.17	1.33
3 years	21.3	(18.7)	8.5	(20.8)	45.0	(42.6)	22.6	(22.7)							
Jenkins Type A Hard Driving Competitiveness Subscale															
Before	N/A	N/A	N/A	N/A	N/A	N/A	N/A	N/A							
3 months	33.5	(38.5)	18.6	(29.2)	7.1	(8.0)	5.3	(9.1)	0.03	2.13	0.09	0.95	11.03**	0.71	2.78
3 years	16.0	(18.6)	10.2	(22.8)	44.7	(34.9)	15.2	(19.0)							
Recent Life Changes Number Endorsed															
Before	N/A	N/A	N/A	N/A	N/A	N/A	N/A	N/A							
3 months	30.5	(40.8)	28.3	(36.9)	9.9	(9.3)	1.0	(1.4)	0.00	0.22	0.04	1.27	13.31**	2.59	0.67
3 years	8.8	(17.5)	17.0	(29.4)	24.4	(30.0)	48.3	(36.4)							

Recent Life Changes Scaled Total															
Before	N/A	N/A	N/A	N/A	N/A	N/A	N/A	N/A							
3 months	30.5	(40.8)	28.3	(36.9)	7.7	(9.8)	5.6	(6.7)	0.00	0.12	0.01	0.71	10.77**	0.79	0.01
3 years	8.8	(17.5)	17.0	(29.4)	28.9	(28.1)	40.4	(38.0)							
Recent Life Changes Subjective Totals															
Before	N/A	N/A	N/A	N/A	N/A	N/A	N/A	N/A							
3 months	11.0	(17.3)	8.0	(9.9)	8.4	(11.1)	5.0	(9.0)	4.45[†]	0.04	0.12	3.18	8.91*	0.41	0.21
3 years	0.0	(0.0)	0.0	(0.0)	37.4	(28.9)	51.3	(43.1)							
Left Main Coronary Disease															
Before	N/A	N/A	N/A	N/A	N/A	N/A	N/A	N/A							
3 months	9.8	(14.2)	31.5	(37.8)	10.7	(9.0)	2.6	(5.6)	0.17	1.27	1.51	1.33	9.53**	0.27	0.55
3 years	0.0	(0.0)	19.2	(22.3)	27.7	(32.5)	34.0	(32.9)							
Age															
Before	N/A	N/A	N/A	N/A	N/A	N/A	N/A	N/A							
3 months	31.7	(37.6)	9.5	(14.4)	3.8	(7.5)	7.4	(8.7)	0.29	2.28	0.66	2.28	11.57**	0.52	0.96
3 years	19.2	(22.2)	0.0	(0.0)	42.3	(39.6)	26.1	(28.5)							
Prior Infarction															
Before	N/A	N/A	N/A	N/A	N/A	N/A	N/A	N/A							
3 months	18.5	(34.4)	25.7	(32.3)	5.5	(6.4)	6.7	(9.2)	0.02	0.07	0.00	1.34	8.31	0.08	0.24
3 years	12.8	(25.5)	10.7	(16.6)	28.5	(47.5)	32.2	(25.3)							

(continued)

Table A-3. (*Continued*)

	Treatment				Control				F-Ratio[a]						
	Low IM		High IM		Low IM		High IM								
Intake Measure	$\bar{X}$	*SD*	$\bar{X}$	*SD*	$\bar{X}$	*SD*	$\bar{X}$	*SD*	G	IM	GIM	T	TG	TIM	TGIM
History of Hypertension															
Before	N/A	N/A	N/A	N/A	N/A	N/A	N/A	N/A							
3 months	43.3	(42.4)	9.2	(11.9)	10.5	(9.2)	2.7	(5.5)	0.02	9.01**	1.39	1.36	10.40**	0.08	0.56
3 years	28.8	(21.3)	0.0	(0.0)	41.7	(41.8)	22.0	(18.0)							
Family History of Heart Disease															
Before	N/A	N/A	N/A	N/A	N/A	N/A	N/A	N/A							
3 months	31.5	(37.8)	9.8	(14.2)	2.0	(5.3)	13.2	(7.5)	0.29	0.42	2.97	1.27	8.98**	0.00	0.07
3 years	19.2	(22.2)	0.0	(0.0)	28.3	(31.4)	35.6	(34.9)							
Triple Coronary Disease															
Before	N/A	N/A	N/A	N/A	N/A	N/A	N/A	N/A							
3 months	18.3	(27.8)	29.5	(39.6)	6.3	(8.9)	6.3	(7.5)	0.03	0.02	1.29	0.57	7.80**	1.37	0.72
3 years	8.5	(20.8)	16.0	(18.6)	38.2	(34.4)	15.0	(17.8)							
Number of Months with Angina															
Before	N/A	N/A	N/A	N/A	N/A	N/A	N/A	N/A							
3 months	21.0	(28.1)	24.6	(37.7)	4.3	(7.5)	10.4	(9.0)	0.11	0.01	0.03	1.60	10.39**	1.08	0.00
3 years	16.0	(23.2)	7.0	(15.7)	35.9	(38.4)	30.6	(22.1)							
Number of Bypasses															
Before	N/A	N/A	N/A	N/A	N/A	N/A	N/A	N/A							
3 months	29.5	(39.6)	18.3	(27.8)	4.6	(6.0)	10.3	(11.9)	0.39	0.80	3.97†	3.88†	17.89***	3.98†	2.69
3 years	16.0	(18.6)	8.5	(20.8)	17.8	(18.4)	61.0	(36.7)							

Systolic Blood Pressure															
Before	N/A	N/A	N/A	N/A	N/A	N/A	N/A	N/A							
3 months	32.0	(37.3)	9.0	(14.7)	5.0	(8.3)	7.8	(8.6)	0.16	2.63	0.70	1.37	9.74**	0.46	1.05
3 years	19.2	(22.2)	0.0	(0.0)	38.6	(37.9)	22.3	(22.3)							
Diastolic Blood Pressure															
Before	N/A	N/A	N/A	N/A	N/A	N/A	N/A	N/A							
3 months	18.2	(29.5)	34.3	(37.6)	8.4	(11.1)	5.0	(6.3)	0.00	0.16	0.58	0.90	9.56**	0.18	0.17
3 years	10.2	(22.8)	16.0	(18.6)	33.2	(33.1)	29.8	(32.8)							
Ejection Fraction															
Before	N/A	N/A	N/A	N/A	N/A	N/A	N/A	N/A							
3 months	28.7	(36.5)	23.7	(33.4)	6.5	(7.2)	7.3	(10.3)	0.08	0.16	0.01	2.19	11.01**	0.11	0.06
3 years	17.0	(29.4)	10.7	(16.6)	42.5	(41.4)	35.7	(29.1)							
New York Heart Association Functional Class															
PRE	III		IV		III		IV								
Before	N/A	N/A	N/A	N/A	N/A	N/A	N/A	N/A							
3 months	31.0	(40.3)	25.8	(30.0)	8.8	(10.2)	5.6	(6.7)	0.01	0.10	0.00	0.75	8.54**	1.25	0.01
3 years	8.9	(17.5)	20.0	(24.8)	27.7	(31.5)	38.4	(39.9)							
MBHI Introversive															
After															
3 months	42.5	(38.9)	9.25	(14.5)	5.4	(6.6)	4.8	(9.5)	0.10	2.44	1.72	0.33	5.06*	0.02	0.13
3 years	25.5	(36.1)	0.00	(0.0)	29.4	(34.1)	25.0	(37.2)							
MBHI Inhibited															
After															
3 months	16.0	(14.5)	24.7	(39.3)	1.6	(3.7)	11.5	(8.2)	0.40	4.73*	0.55	1.34	8.71**	2.39	0.83
3 years	0.0	(0.0)	17.0	(29.4)	12.4	(11.5)	54.8	(44.0)							
MBHI Cooperative															
After															
3 months	15.7	(15.0)	25.0	(39.0)	6.2	(8.8)	4.3	(6.5)	0.08	0.01	1.95	0.77	6.83*	0.45	1.52
3 years	0.0	(0.0)	17.0	(29.4)	43.0	(43.7)	15.2	(16.9)							

(continued)

Table A-3. *(Continued)*

Intake Measure	Treatment Low IM $\bar{X}$	Treatment Low IM *SD*	Treatment High IM $\bar{X}$	Treatment High IM *SD*	Control Low IM $\bar{X}$	Control Low IM *SD*	Control High IM $\bar{X}$	Control High IM *SD*	*F*-Ratio[a] G	IM	GIM	T	TG	TIM	TGIM
MBHI Sociable															
After															
3 months	34.0	(34.6)	6.7	(7.2)	11.3	(10.0)	2.9	(5.1)	1.91	15.47**	0.81	5.04*	20.77***	4.48*	9.76**
3 years	17.0	(29.4)	0.0	(0.0)	73.0	(30.1)	10.9	(11.5)							
MBHI Confident															
After															
3 months	18.2	(29.5)	31.0	(0.0)	11.7	(9.5)	2.8	(5.1)	0.24	0.54	0.70	0.21	7.12*	1.62	0.00
3 years	10.2	(22.8)	0.0	(0.0)	50.3	(26.1)	19.4	(33.2)							
MBHI Forceful															
After															
3 months	19.0	(34.0)	23.0	(11.3)	4.0	(7.3)	5.9	(7.8)	0.02	0.02	0.30	0.18	5.47*	0.02	0.89
3 years	12.8	(25.5)	0.0	(0.0)	19.0	(19.5)	32.9	(39.9)							
MBHI Respectful															
After															
3 months	5.3	(6.6)	50.5	(27.6)	0.0	(0.0)	6.3	(7.7)	0.76	6.43*	0.97	0.01	3.80†	0.00	1.37
3 years	0.0	(0.0)	25.5	(36.1)	7.5	(10.6)	32.3	(35.4)							
MBHI Sensitive															
After															
3 months	9.3	(14.5)	42.5	(38.9)	2.9	(7.1)	9.3	(6.5)	0.03	3.59†	1.07	0.42	5.39*	0.01	0.15
3 years	0.0	(0.0)	25.5	(36.1)	23.9	(26.7)	34.8	(46.8)							
MBHI Chronic Tension															
After															
3 months	5.25	(6.6)	50.5	(27.6)	0.25	(0.5)	8.0	(8.0)	0.45	8.19**	1.21	0.13	5.69*	0.02	1.43
3 years	0.0	(0.0)	25.5	(36.1)	12.8	(13.7)	36.4	(39.1)							
MBHI Recent Stress															
After															
3 months	10.4	(12.8)	70.0	(0.0)	5.0	(7.9)	5.4	(7.5)	2.63	9.85**	6.01*	0.22	4.42†	0.01	0.35
3 years	0.0	(0.0)	51.0	(0.0)	21.8	(30.8)	35.0	(38.5)							

MBHI Premorbid Pessimism															
After															
3 months	10.4	(12.8)	70.0	(0.0)	4.7	(7.2)	7.5	(10.6)	0.80	22.14***	2.65	2.08	10.69**	1.51	3.10
3 years	0.0	(0.0)	51.0	(0.0)	18.6	(25.4)	69.5	(41.7)							
MBHI Future Dispair															
After															
3 months	2.0	(1.0)	38.7	(28.3)	3.3	(6.7)	13.5	(2.1)	0.04	1.55	1.48	0.08	3.21†	1.57	0.00
3 years	0.0	(0.0)	17.0	(29.4)	29.6	(35.7)	20.0	(28.3)							
MBHI Social Alienation															
After															
3 months	2.5	(0.7)	29.3	(29.8)	3.3	(5.3)	8.5	(9.9)	1.03	6.85*	0.15	2.60	10.39**	1.51	5.84*
3 years	0.0	(0.0)	12.8	(25.5)	10.3	(12.3)	58.5	(38.0)							
MBHI Somatic Anxiety															
After															
3 months	6.7	(7.2)	34.0	(34.6)	2.9	(7.1)	9.3	(6.5)	0.10	2.33	0.45	0.58	5.45*	0.04	0.24
3 years	0.0	(0.0)	17.0	(29.4)	23.9	(26.7)	34.8	(46.8)							
MBHI Allergic Inclination															
After															
3 months	5.3	(6.6)	50.5	(27.6)	0.2	(0.4)	11.2	(7.1)	0.14	4.76*	2.44	0.22	6.18*	1.01	0.10
3 years	0.0	(0.0)	25.5	(36.1)	27.5	(36.6)	28.2	(33.4)							
MBHI Gastrointestinal Susceptibility															
After															
3 months	6.7	(7.2)	34.0	(34.6)	2.9	(7.1)	9.3	(6.5)	0.10	2.33	0.45	0.58	5.45*	0.04	0.24
3 years	0.0	(0.0)	17.0	(29.4)	23.9	(26.7)	34.8	(46.8)							

(*continued*)

Table A-3. (*Continued*)

	Treatment				Control				*F*-Ratio[a]						
	Low IM		High IM		Low IM		High IM								
Intake Measure	$\bar{X}$	*SD*	$\bar{X}$	*SD*	$\bar{X}$	*SD*	$\bar{X}$	*SD*	G	IM	GIM	T	TG	TIM	TGIM
MBHI Cardiovascular Tendency															
After															
3 months	5.3	(6.6)	50.5	(27.6)	0.2	(0.4)	9.3	(7.8)	0.30	9.82**	1.05	0.21	6.41*	0.01	1.64
3 years	0.0	(0.0)	25.5	(36.1)	13.2	(11.9)	40.0	(41.6)							
MBHI Pain Proneness															
After															
3 months	12.5	(13.8)	36.0	(48.1)	4.6	(7.2)	8.0	(9.9)	0.00	1.81	0.62	0.58	4.42†	0.05	0.01
3 years	0.0	(0.0)	25.5	(36.1)	26.1	(37.0)	35.5	(6.4)							
MBHI Life Threat Reactivity															
After															
3 months	6.7	(7.2)	34.0	(34.6)	3.8	(7.1)	9.0	(7.9)	0.31	3.48†	0.11	1.07	6.89*	0.11	1.05
3 years	0.0	(0.0)	17.0	(29.4)	20.9	(26.2)	46.3	(49.8)							
MBHI Emotional Vulnerability															
After															
3 months	11.3	(17.0)	29.3	(35.7)	4.8	(9.5)	5.4	(6.6)	0.03	0.80	0.58	0.49	5.11*	0.00	0.01
3 years	0.0	(0.0)	17.0	(29.4)	26.5	(36.8)	28.6	(34.4)							

[a]G = Group (treatment versus control), T = time, IM = Intake Measure.
†$p < .10$. *$p < .05$. **$p < .01$. ***$p < .001$.

pressed. However, the increase in depression for control patients with many bypasses had a much larger change than the other three cells. The Sociable Personality Style on the MBHI had both a significant interaction with time and a significant triple interaction with Group and time. Those patients low on the Sociable scale improved more in the treatment group but declined more in the control group. Also, those patients high on Sociable Alienation improved more in the treatment group but declined more in the control group.

Table A-3 presents the Analogue results at 3 months factorially crossed with Group and Intake Measure. GSI had a main effect with low GSI scores before surgery associated with higher report of depression 3 months after surgery. On the Jenkins Speed and Impatience Subscale Type B patients were reported to have less postsurgical depression. The Jenkins Total Scale had a similar effect but only to a marginally significant extent. Number of Bypasses also had a marginally significant main effect such that patients with fewer bypasses were reported to have less depression than those with many bypasses. On the MBHI, a low score on the Recent Stress, Future Despair, Pain Proneness, Life Threat Reactivity, and to a marginal extent Premorbid Pessimism, Somatic Anxiety, and Gastrointestinal Susceptibility scales was associated with a low score on the Analogue Scale. Finally, History of Hypertension interacted significantly with Group. Those with such a history were reported to have more depression than those without such a history in the treatment group. However, those with this history were reported to have less depression than those without it in the control group.

Thus, there were many individual presurgical variables with effects on later postsurgical depression. Appendix B presents analyses that attempt to integrate these variables.

Table A-4. Means, Standard Deviations, and *F*-Ratios of Group by Various Intake Measures for the Analogues at 3 Months after Surgery

	Treatment				Control										
	Low IM		High IM		Low IM		High IM		*F*-Ratio[a]						
Intake Measure	$\bar{X}$	*SD*	$\bar{X}$	*SD*	$\bar{X}$	*SD*	$\bar{X}$	*SD*	G	IM	GIM	T	TG	TIM	TGIM
Locus of Control															
Before	N/A	N/A	N/A	N/A	N/A	N/A	N/A	N/A							
3 months	12.8	(28.6)	16.0	(20.9)	14.4	(29.0)	24.0	(33.2)	0.35	0.63	0.16	N/A	N/A	N/A	N/A
SCL-90 Global Severity Index															
Before	N/A	N/A	N/A	N/A	N/A	N/A	N/A	N/A							
3 months	13.3	(25.0)	19.3	(23.5)	8.2	(11.1)	31.7	(38.2)	0.26	4.23*	1.48	N/A	N/A	N/A	N/A
SCL-90 Positive Symptom Total															
Before	N/A	N/A	N/A	N/A	N/A	N/A	N/A	N/A							
3 months	15.4	(26.5)	16.6	(22.7)	9.6	(11.4)	27.5	(37.1)	0.11	1.67	1.27	N/A	N/A	N/A	N/A
Concept Level Analogy Test															
Before	N/A	N/A	N/A	N/A	N/A	N/A	N/A	N/A							
3 months	28.3	(27.5)	9.9	(15.3)	21.4	(32.7)	11.4	(22.1)	0.10	2.61	0.23	N/A	N/A	N/A	N/A
Jenkins Type A Overall Score															
Before	N/A	N/A	N/A	N/A	N/A	N/A	N/A	N/A							
3 months	31.7	(31.8)	5.0	(5.8)	17.6	(34.0)	16.0	(17.0)	0.04	3.00[†]	2.36	N/A	N/A	N/A	N/A
Jenkins Type A Speed Impatience Subscale															
Before	N/A	N/A	N/A	N/A	N/A	N/A	N/A	N/A							
3 months	30.7	(29.8)	6.1	(10.4)	21.5	(32.6)	11.3	(18.5)	0.06	4.81*	0.82	N/A	N/A	N/A	N/A

Jenkins Type A Job Involvement Subscale															
Before	N/A	N/A	N/A	N/A	N/A	N/A	N/A	N/A							
3 months	24.0	(35.2)	16.8	(22.0)	25.3	(38.8)	13.0	(20.9)	0.02	1.04	0.07	N/A	N/A	N/A	N/A
Jenkins Type A Hard Driving Competitiveness Subscale															
Before	N/A	N/A	N/A	N/A	N/A	N/A	N/A	N/A							
3 months	21.3	(28.9)	17.1	(25.3)	21.2	(32.9)	11.7	(18.1)	0.10	0.60	0.09	N/A	N/A	N/A	N/A
Recent Life Changes Number Endorsed															
Before	N/A	N/A	N/A	N/A	N/A	N/A	N/A	N/A							
3 months	23.7	(32.8)	15.9	(24.2)	15.9	(27.3)	29.3	(39.0)	0.07	0.07	1.05	N/A	N/A	N/A	N/A
Recent Life Changes Scaled Total															
Before	N/A	N/A	N/A	N/A	N/A	N/A	N/A	N/A							
3 months	23.7	(32.8)	15.9	(24.2)	14.5	(22.7)	28.6	(41.0)	0.03	0.09	1.14	N/A	N/A	N/A	N/A
Recent Life Changes Subjective Totals															
Before	N/A	N/A	N/A	N/A	N/A	N/A	N/A	N/A							
3 months	7.2	(13.4)	14.0	(17.6)	8.1	(10.1)	12.5	(23.0)	0.00	0.70	0.03	N/A	N/A	N/A	N/A
Left Main Coronary Disease															
Before	N/A	N/A	N/A	N/A	N/A	N/A	N/A	N/A							
3 months	12.5	(16.8)	18.2	(24.1)	11.2	(9.5)	22.5	(34.7)	0.05	1.62	0.17	N/A	N/A	N/A	N/A

(continued)

Table A-4. (*Continued*)

	Treatment				Control				*F*-Ratio[a]						
	Low IM		High IM		Low IM		High IM								
Intake Measure	$\bar{X}$	*SD*	$\bar{X}$	*SD*	$\bar{X}$	*SD*	$\bar{X}$	*SD*	G	IM	GIM	T	TG	TIM	TGIM
Age															
Before	N/A	N/A	N/A	N/A	N/A	N/A	N/A	N/A							
3 months	16.4	(24.6)	13.4	(16.1)	24.0	(37.1)	12.9	(19.7)	0.36	1.38	0.44	N/A	N/A	N/A	N/A
Prior Infarction															
Before	N/A	N/A	N/A	N/A	N/A	N/A	N/A	N/A							
3 months	14.5	(21.9)	18.3	(22.1)	17.2	(30.8)	19.8	(29.3)	0.11	0.24	0.01	N/A	N/A	N/A	N/A
History of Hypertension															
Before	N/A	N/A	N/A	N/A	N/A	N/A	N/A	N/A							
3 months	28.7	(27.8)	9.0	(13.3)	11.3	(16.7)	24.6	(35.4)	0.02	0.27	7.25**	N/A	N/A	N/A	N/A
Family History of Heart Disease															
Before	N/A	N/A	N/A	N/A	N/A	N/A	N/A	N/A							
3 months	17.7	(24.5)	11.4	(16.1)	23.4	(35.5)	12.9	(8.2)	0.27	1.46	0.09	N/A	N/A	N/A	N/A
Triple Coronary Disease															
Before	N/A	N/A	N/A	N/A	N/A	N/A	N/A	N/A							
3 months	15.6	(20.3)	17.3	(24.6)	19.9	(32.7)	16.3	(19.6)	0.06	0.02	0.16	N/A	N/A	N/A	N/A
Number of Months with Angina															
Before	N/A	N/A	N/A	N/A	N/A	N/A	N/A	N/A							
3 months	12.4	(19.0)	21.1	(24.6)	17.2	(25.5)	21.8	(33.8)	0.18	1.07	0.10	N/A	N/A	N/A	N/A
Number of Bypasses															
Before	N/A	N/A	N/A	N/A	N/A	N/A	N/A	N/A							
3 months	13.9	(20.7)	22.1	(24.4)	13.2	(22.4)	29.1	(38.0)	0.22	3.31†	0.34	N/A	N/A	N/A	N/A

Systolic Blood Pressure															
Before	N/A	N/A	N/A	N/A	N/A	N/A	N/A	N/A							
3 months	19.0	(25.6)	12.8	(15.9)	14.8	(26.7)	22.9	(32.1)	0.21	0.02	1.29	N/A	N/A	N/A	N/A
Diastolic Blood Pressure															
Before	N/A	N/A	N/A	N/A	N/A	N/A	N/A	N/A							
3 months	15.0	(21.4)	18.5	(23.0)	22.5	(33.2)	16.1	(26.8)	0.16	0.05	0.58	N/A	N/A	N/A	N/A
Ejection Fraction															
Before	N/A	N/A	N/A	N/A	N/A	N/A	N/A	N/A							
3 months	16.1	(22.4)	17.6	(23.4)	28.9	(36.1)	8.3	(9.7)	0.07	2.10	2.81	N/A	N/A	N/A	N/A
New York Heart Association Functional Class															
Before	N/A	N/A	N/A	N/A	N/A	N/A	N/A	N/A							
3 months	21.9	(23.1)	17.3	(22.8)	15.9	(31.7)	16.5	(27.6)	0.54	1.10	0.09	N/A	N/A	N/A	N/A
MBHI Introversive															
Before	N/A	N/A	N/A	N/A	N/A	N/A	N/A	N/A							
3 months	25.2	(30.1)	8.3	(13.2)	16.8	(23.6)	19.0	(36.2)	0.01	0.63	1.07	N/A	N/A	N/A	N/A
MBHI Inhibited															
Before	N/A	N/A	N/A	N/A	N/A	N/A	N/A	N/A							
3 months	13.1	(16.7)	15.6	(26.6)	11.4	(29.5)	24.3	(24.8)	0.16	0.76	0.36	N/A	N/A	N/A	N/A
MBHI Cooperative															
Before	N/A	N/A	N/A	N/A	N/A	N/A	N/A	N/A							
3 months	12.6	(17.1)	16.1	(26.2)	21.6	(32.6)	12.1	(19.3)	0.08	0.11	0.53	N/A	N/A	N/A	N/A
MBHI Sociable															
Before	N/A	N/A	N/A	N/A	N/A	N/A	N/A	N/A							
3 months	22.3	(27.9)	8.4	(14.0)	15.5	(8.9)	18.3	(32.4)	0.03	0.36	0.82	N/A	N/A	N/A	N/A
MBHI Confident															
Before	N/A	N/A	N/A	N/A	N/A	N/A	N/A	N/A							
3 months	14.3	(23.3)	14.4	(20.0)	17.1	(17.9)	17.8	(33.8)	0.11	0.00	0.00	N/A	N/A	N/A	N/A

(continued)

Table A-4. (*Continued*)

Intake Measure	Treatment				Control				*F*-Ratio[a]						
	Low IM		High IM		Low IM		High IM								
	$\bar{X}$	*SD*	$\bar{X}$	*SD*	$\bar{X}$	*SD*	$\bar{X}$	*SD*	G	IM	GIM	T	TG	TIM	TGIM
MBHI Forceful															
Before	N/A	N/A	N/A	N/A	N/A	N/A	N/A	N/A							
3 months	14.3	(24.9)	14.5	(17.9)	24.8	(36.1)	13.1	(21.1)	0.25	0.40	0.44	N/A	N/A	N/A	N/A
MBHI Respectful															
Before	N/A	N/A	N/A	N/A	N/A	N/A	N/A	N/A							
3 months	10.0	(14.8)	25.3	(33.2)	23.3	(32.4)	14.6	(25.5)	0.02	0.12	1.57	N/A	N/A	N/A	N/A
MBHI Sensitive															
Before	N/A	N/A	N/A	N/A	N/A	N/A	N/A	N/A							
3 months	12.9	(16.6)	17.0	(30.3)	13.4	(30.9)	21.3	(24.9)	0.07	0.43	0.04	N/A	N/A	N/A	N/A
MBHI Chronic Tension															
Before	N/A	N/A	N/A	N/A	N/A	N/A	N/A	N/A							
3 months	11.1	(15.2)	20.2	(30.9)	12.4	(20.7)	20.7	(31.3)	0.01	0.90	0.00	N/A	N/A	N/A	N/A
MBHI Recent Stress															
Before	N/A	N/A	N/A	N/A	N/A	N/A	N/A	N/A							
3 months	8.2	(12.3)	37.0	(35.2)	5.6	(7.9)	24.8	(32.7)	0.62	6.62*	0.26	N/A	N/A	N/A	N/A
MBHI Premorbid Pessimism															
Before	N/A	N/A	N/A	N/A	N/A	N/A	N/A	N/A							
3 months	11.9	(15.4)	23.3	(40.4)	5.2	(7.3)	31.1	(35.1)	0.00	4.08†	0.61	N/A	N/A	N/A	N/A
MBHI Future Dispair															
Before	N/A	N/A	N/A	N/A	N/A	N/A	N/A	N/A							
3 months	7.6	(14.8)	21.1	(25.8)	7.9	(16.4)	33.1	(35.3)	0.56	5.54*	0.56	N/A	N/A	N/A	N/A
MBHI Social Alienation															
Before	N/A	N/A	N/A	N/A	N/A	N/A	N/A	N/A							
3 months	8.7	(15.9)	18.6	(24.9)	16.3	(30.8)	19.2	(24.0)	0.21	0.52	0.15	N/A	N/A	N/A	N/A

MBHI Somatic Anxiety															
Before	N/A	N/A	N/A	N/A	N/A	N/A	N/A	N/A							
3 months	11.0	(15.3)	20.4	(30.7)	3.9	(7.5)	27.8	(32.7)	0.00	3.80†	0.72	N/A	N/A	N/A	N/A
MBHI Allergic Inclination															
Before	N/A	N/A	N/A	N/A	N/A	N/A	N/A	N/A							
3 months	11.1	(15.2)	20.2	(30.9)	7.4	(18.4)	26.7	(31.8)	0.03	2.65	0.34	N/A	N/A	N/A	N/A
MBHI Gastrointestinal Susceptibility															
Before	N/A	N/A	N/A	N/A	N/A	N/A	N/A	N/A							
3 months	11.0	(15.3)	20.4	(30.7)	3.9	(7.5)	27.8	(32.7)	0.00	3.80†	0.72	N/A	N/A	N/A	N/A
MBHI Cardiovascular Tendency															
Before	N/A	N/A	N/A	N/A	N/A	N/A	N/A	N/A							
3 months	10.0	(14.8)	25.3	(33.2)	17.3	(34.0)	17.7	(21.7)	0.00	0.69	0.62	N/A	N/A	N/A	N/A
MBHI Pain Proneness															
Before	N/A	N/A	N/A	N/A	N/A	N/A	N/A	N/A							
3 months	10.6	(13.8)	19.3	(29.5)	4.8	(7.1)	34.4	(35.4)	0.34	5.75*	1.71	N/A	N/A	N/A	N/A
MBHI Life Threat Reactivity															
Before	N/A	N/A	N/A	N/A	N/A	N/A	N/A	N/A							
3 months	9.6	(14.7)	19.1	(26.8)	3.9	(6.8)	35.7	(34.3)	0.49	7.09*	2.04	N/A	N/A	N/A	N/A
MBHI Emotional Vulnerability															
Before	N/A	N/A	N/A	N/A	N/A	N/A	N/A	N/A							
3 months	13.0	(18.1)	15.4	(24.7)	11.5	(20.3)	21.2	(31.3)	0.06	0.45	0.17	N/A	N/A	N/A	N/A

[a]G = Group (treatment versus control), T = time, IM = Intake Measure.
$^{\dagger}p < .10$. $^{*}p < .05$. $^{**}p < .01$.

B

Presurgical Variables Predictive of Depression

This appendix examines the effects of combinations of presurgical variables on depression after surgery. It also establishes how predictive of later depression these variables are. In order to identify presurgical variables predictive of postsurgical depression, stepwise multiple regression equations were calculated. Potential independent variables were selected on the basis of the ANOVAs reviewed in Appendix A. Because the MBHI is a relatively new instrument and its scales less well known, it was not included in these analyses. Among the other instruments, intake (i.e., presurgical) measures that had at least marginally significant main effects or marginally significant interactions with time were included as potential independent variables for the regression analyses. Also included as potential independent variables were interactional variables created by multiplying the grouping variable (treatment = 1, control = 2) by those intake measures which had at least marginally significant triple interactions (i.e., time by Group by Intake Measure) or *both* time by Intake Measure *and* Group by Intake interactions. This pool of potential independent variables was then selected in a stepwise fashion for inclusion in the regression equations (Jenkins, Sours, & Babines, 1981). The criterion for inclusion was an F-ratio significant at better than the .10 level and the significance for exclusion following inclusion better than .125. Four separate regression equations were calculated: one each for BDI at 3 years, BDI at 3 months, Analogue at 3 years, and Analogue at 3 months.

Table B-1 presents the very significant multiple regression for the BDI at 3 years. As can be seen from inspection of Table B-1, five independent variables were allowed to enter the equation all with significant regression coefficients. These variables and their direction are as follows: (1) Group by RLCNE: being in the control group and having many life changes tended to increase later depression; (2) Triple Coronary Disease: having this disease tended to increase later depression (note that the coding of this and other dichotomous medical items will lead to a negative coefficient if the item's presence tends to increase depression); (3) GSI: increases in the SCL-90 Global Severity Index tended to increase postsurgical depression; (4) New York Heart Association Functional Class: higher levels on this scale tended to lead to higher levels of later depres-

Table B-1. Regression Analysis of Beck Depression Inventory at 3 Years

Equation statistics
(Beck Depression Inventory at 3 years is dependent variable) $R = .95$, $R^2 = .91$, $R^2_{Adj.} = .87$, $F(5,12) = 23.37$, $p = .001$

Independent variable statistics

Variable	Standardized regression coefficient	Standard error of standard regression coefficient	F	p
Group by Recent Life Changes Number Endorsed	.79	.11	55.14	.0001
Triple Coronary Disease	−.34	.09	14.71	.0024
SCL-90 Global Severity Index before surgery	.24	.10	5.79	.0331
New York Heart Association Functional Class prior to surgery	.28	.10	7.62	.0172
Age	.30	.11	8.00	.0152

sion; and (5) Age: older patients tended to have higher levels of later depression. Together these variables accounted for more than 90% of the variance in this sample for BDI measured depression at 3 years after surgery.

The equation for BDI at 3 months, presented in Table B-2, was also highly significant. Four independent variables were selected for inclusion in this equation, all with very significant regression coefficients. The direction of these four variables was as follows: (1) BDI: high presurgical BDI scores tended to be associated with high BDI scores at 3 months after surgery; (2) Group by Number of Bypasses: having a greater number of bypasses and being in the control group tended to result in higher levels of later depression; (3) Group by RLCNE: the greater the number of recent life changes and being in the control group tended to lead to greater postsurgical depression; and (4) Group by Left Main Coronary Disease: being in the treatment group and having left main coronary disease tended to reduce postsurgical depression. Together these four variables account for 80% of the variance in this sample on depression at 3 months after surgery as measured by the BDI.

Table B-3 presents the results of the regression analysis of the Analogue Depression Scale at 3 years. Although the multiple R is not quite so large and the corresponding F-ratio not quite so significant as in the other three regression

Table B-2. Regression Analysis of Beck Depression Inventory at 3 Months

Equation statistics
(Beck Depression Inventory at 3 months is dependent variable) $R = .90$, $R^2 = .80$, $R^2_{Adj.} = .77$, $F(4,23) = 23.70$, $p = .0001$

	Independent variable statistics			
Variable	Standardized regression coefficient	Standard error of standard regression coefficient	F	p
Beck Depression Inventory before surgery	.51	.12	17.42	.0004
Group by Number of Bypasses	.37	.10	13.22	.0014
Group by Recent Life Change Number Endorsed before surgery	.33	.12	7.31	.0127
Group by Left Main Coronary Disease	−.26	.10	6.50	.0180

equations, both statistics are nevertheless quite strong. Only three variables were selected for this equation. An examination of Table B-3 reveals that the first two variables selected have significant regression coefficients and the third one was marginally significant. These variables and their direction are as follows: (1) Group by RLCNE: having more recent life changes and being in the control group tended to lead to greater reported depression at 3 years after surgery; (2) Number of Bypasses: the greater the number of bypasses, the greater the tendency to have later depression; and (3) History of Hypertension: having a history of hypertension tended to be associated with greater reported depression later. In this sample, this combination of variables accounted for about two-thirds of the variance in reported depression 3 years after surgery.

The highly significant results of the regression analysis of the Analogue Depression Scale at 3 months are presented in Table B-4. Eight variables were included in this equation with six of the first seven having significant regression coefficients and the other two having a marginally significant regression coefficient. The direction of these variables is as follows: (1) GSI: the higher the SCL-90 Global Severity Index, the greater the chance for reports of elevated postsurgical depression; (2) Group by GSI: being in treatment and having a low SCL-90 Global Severity Index tends to lead to reports of higher postsurgical

Table B-3. Regression Analysis of Analogue Depression at 3 Years

Equation statistics
(Analogue Depression Scale at 3 years is dependent variable) $R = .82$, $R^2 = .68$, $R^2_{Adj.} = .60$, $F(3,12) = 8.42$, $p = .003$

Independent variable statistics

Variable	Standardized regression coefficient	Standard error of standard regression coefficient	F	p
Group by Recent Life Change Number Endorsed before surgery	.54	.17	10.69	.0067
Number of Bypasses	.39	.17	5.28	.0404
History of Hypertension	−.34	.17	4.11	.0655

depression; (3) Number of Bypasses: having had more bypasses is associated with reports of higher postsurgical depression; (4) LOC: the more internal the patient is, the more likely that he will be reported to have greater depression later; (5) RLCNE: the more recent life changes before surgery, the greater the reports of depression after surgery; (6) Age: the older the patient, the greater his reported postsurgical depression; (7) Group by Left Main Coronary Disease: being in the treatment group and having left main coronary disease is associated with later reports of *less* depression; (8) Triple Coronary Disease: having this disease tends to increase postsurgical reports of later depression. These eight variables combined in this fashion account for 86% of the variance in the Analogue Depression Scale 3 months after surgery. Given the relatively large number of variables in this equation, however, it may not generalize well to other samples.

The purpose of this section was to determine if and in what way presurgical variables could predict later postsurgical depression. The large mutiple Rs and highly significant F-ratios in these four equations suggest that a very strong relationship existed in this sample. The fact that the independent measures making up each of the equations vary a great deal, however, suggests that the prediction is highly complex. No variable appears in all four equations. Only Group by Number of Recent Life Changes (being in the control group and having many life changes associated with later depression) appears in three equations, and it is the strongest independent variable in two of them. Several variables appeared in two equations. They are SCL-90 Global Severity Index (a

Table B-4. Regression Analysis of Analogue Depression at 3 Months

Equation statistics
(Analogue Depression Scale at 3 months is dependent variable) $R = .93$, $R^2 = .86$, $R^2_{Adj.} = .77$, $F(8,12) = 9.49$, $p = .0004$

Independent variable statistics

Variable	Standardized regression coefficient	Standard error of standard regression coefficient	F	p
SCL-90 Global Severity Index before surgery	1.34	.22	37.06	.0001
Group by SCL-90 Global Severity Index before surgery	−1.53	.31	24.53	.0003
Number of Bypasses	.28	.16	3.27	.0958
Locus of Control External before surgery	−.59	.17	12.40	.0042
Recent Life Change Number Endorsed before surgery	.43	.15	8.70	.0122
Age	.82	.25	10.53	.0070
Group by Left Main Coronary Disease	.51	.21	5.98	.0309
Triple Coronary Disease	−.29	.14	4.08	.0664

higher score tending toward more depression), Age (older tending toward more depression), Triple Coronary Disease (having the disease tending toward more depression), and Number of Bypasses (the more bypasses the greater the depression). Group by Left Main Disease also appeared in two equations but had contradictory signs. Noticeable for its absence is the Jenkins. Evidently Type A did not greatly affect of depression in this sample. Thus, although the impact of presurgical psychological and medical state on postsurgical depression is complex, it is also very powerful. Future research and clinical practice should carefully attend to the variables outlined above.

REFERENCE

Jenkins, J., Sours, K., & Babines, T. New regression. In C. H. Hull & N. H. Nie (Eds.), *Statistical package for social science—Version 9*. New York: McGraw-Hill, 1981.

C

Correlations between the MBHI and Other Measures

Since the MBHI is a relatively new instrument, its relationship to other better known psychological inventories and medical measures is of interest. Table C-1 presents this information for the sample in this study in the form of Pearson Product Moment Correlations.

Table C-1. Correlations between MBHI Scales and Other Instruments MBHI before Surgery

	Introversion	Inhibited	Cooperative	Sociable	Confident	Forceful	Respectful	Sensitive	Chronic tension	Recent stress	Premorbid pessimism
CLAT											
Before	.02	−.35*	.22	.29	.06	−.22	−.15	−.15	−.29	.01	−.20
3 months	−.05	−.52***	.43**	.66***	.24	−.36*	−.15	−.36*	−.33*	−.06	−.43*
3 years	—	—	—	—	—	—	—	—	—	—	—
GSI											
Before	−.51***	.49***	−.09	−.38**	−.40***	.11	−.27	.57***	.36**	.47***	.54***
3 months	−.33**	.54***	−.04	−.35*	−.44**	−.01	−.27	.53***	.25	.33*	.44**
3 years	−.37	.61**	−.18	−.72***	−.38	.06	−.17	.61**	.49*	.51*	.62**
PST											
Before	−.55***	.51***	.08	−.29*	−.47***	.01	−.24	.58***	.28	.45***	.51***
3 months	−.32*	.42**	.01	−.26	−.36*	−.04	−.15	.45**	.28	.35*	.37*
3 years	−.36	.62**	−.10	−.63**	−.45*	.02	−.31	.58**	.37	.44*	.55**
MMPID											
Before	−.38**	.63***	.09	−.55***	−.59***	−.07	−.06	.53***	.30*	.37*	.58***
3 months	−.36*	.69***	.08	−.54***	−.59***	−.06	.02	.62***	.28	.35*	.59***
3 years	—	—	—	—	—	—	—	—	—	—	—
Beck											
Before	−.52***	.61***	−.11	−.48***	−.47***	.15	−.33*	.66***	.36*	.52***	.65***
3 months	−.26	.61***	−.05	−.49***	−.54***	−.02	−.24	.48***	.23	.49***	.58***
3 years	−.46*	.67***	−.30	−.67***	−.39	.20	−.25	.65***	.49*	.64***	.72***
Locus of Control											
Before	−.28	.59***	−.07	−.47**	−.47***	.13	−.24	.60***	.32*	.34*	.52***
3 months	−.27	.63***	−.21	−.58***	−.44**	.17	−.12	.65***	.41*	.20	.58***
3 years	−.31	.62**	−.03	−.81***	−.51**	−.04	−.05	.61**	.51*	.52*	.67***
Analogue Depression											
Before	—	—	—	—	—	—	—	—	—	—	—
3 months	−.20	.24	−.03	−.20	−.10	−.01	.07	.17	.20	.37*	.36*
3 years	−.40	.44*	−.34	−.46*	−.16	.24	.05	.44*	.56**	.50*	.47*

RLCNE											
Before	−.48**	.20	.12	−.18	−.33*	−.11	−.20	.32	.11	.59***	.40*
3 months	−.22	.03	.03	−.13	−.08	−.13	.12	.16	.21	.39*	.32
3 years	—	—	—	—	—	—	—	—	—	—	—
RLCSCT											
Before	−.45**	.17	.06	−.14	−.30	−.02	−.24	.29	.11	.54***	.34*
3 months	−.14	−.12	.00	.04	.08	−.05	.06	.03	.11	.25	.10
3 years	—	—	—	—	—	—	—	—	—	—	—
RLCSUT											
Before	−.48**	.07	−.07	.09	.04	.18	−.17	.37	.33	.62***	.28
3 months	−.03	−.02	−.06	.00	.04	−.07	.13	.07	.37*	.48**	.28
3 years	—	—	—	—	—	—	—	—	—	—	—
Jenkins Total											
Before	−.26	.09	−.55***	−.07	.39*	.55***	.07	.20	.59***	.27	.14
3 months	—	—	—	—	—	—	—	—	—	—	—
3 years	−.44*	.28	−.48*	−.42	.05	.45*	.30	.45*	.71***	.36	.33
Months with Angina											
Before	.15	.12	−.34**	−.18	.08	.20	.01	.08	.07	.02	.21
3 months	—	—	—	—	—	—	—	—	—	—	—
3 years	—	—	—	—	—	—	—	—	—	—	—
Systolic Blood Pressure											
Before	−.32*	.31*	−.26	−.19	−.04	.34*	.08	.36*	.26	.28	.27
3 months	—	—	—	—	—	—	—	—	—	—	—
3 years	—	—	—	—	—	—	—	—	—	—	—
Diastolic Blood Pressure											
Before	−.07	.25	−.18	−.24	.03	.23	.32*	.23	.19	.23	.17
3 months	—	—	—	—	—	—	—	—	—	—	—
3 years	—	—	—	—	—	—	—	—	—	—	—

(continued)

Table C-1. (*Continued*)

	Introversion	Inhibited	Cooperative	Sociable	Confident	Forceful	Respectful	Sensitive	Chronic tension	Recent stress	Premorbid pessimism
Ejection Fraction prior to Surgery											
Before	.20	.03	.05	.06	−.17	−.02	−.11	−.17	−.33	−.27	−.24
3 months	—	—	—	—	—	—	—	—	—	—	—
3 years	—	—	—	—	—	—	—	—	—	—	—

	Future despair	Social alienation	Somatic anxiety	Allergic inclination	Gastrointestinal susceptibility	Cardiovascular tendency	Pain proneness	Life threat	Emotional vulnerability
CLAT									
Before	−.33*	−.38*	−.21	−.48***	−.25	−.38*	.01	−.36*	−.29
3 months	−.61***	−.62***	−.41*	−.56***	−.47**	−.50**	−.21	−.62***	−.42**
3 years	—	—	—	—	—	—	—	—	—
GSI									
Before	.46***	.41**	.56***	.37*	.43**	.29	.35*	.35*	.47***
3 months	.44**	.50***	.46**	.37*	.36*	.31*	.22	.35*	.38*
3 years	.56**	.72***	.47*	.41	.49*	.43*	.26	.52*	.43*
PST									
Before	.40**	.36*	.53***	.34*	.37**	.26	.40**	.27	.52***
3 months	.36*	.38*	.42**	.33*	.30*	.30	.17	.28	.32*
3 years	.53*	.70***	.39	.33	.40	.33	.29	.41	.45*
MMPID									
Before	.52***	.48***	.58***	.47***	.53***	.34*	.50***	.47***	.49***
3 months	.55***	.60***	.61***	.56***	.56***	.47**	.47**	.47***	.58***
3 years	—	—	—	—	—	—	—	—	—
Beck									
Before	.57***	.53***	.64***	.45**	.55***	.40**	.46**	.48***	.59***
3 months	.55***	.58***	.42**	.41**	.46**	.38*	.53***	.51***	.49***
3 years	.59**	.72***	.52*	.51*	.59**	.48*	.38	.60**	.53*

Locus of Control									
Before	.53***	.50***	.57***	.49***	.53***	.44**	.41**	.52***	.53***
3 months	.64***	.64***	.60***	.53***	.60***	.47**	.34*	.54***	.54***
3 years	.74***	.74***	.57**	.60**	.54*	.60**	.53*	.64**	.34
Analogue Depression									
Before	—	—	—	—	—	—	—	—	—
3 months	.41*	.27	.28	.23	.31	.22	.39*	.38*	.15
3 years	.27	.39	.36	.34	.38	.47*	.14	.37	.38
RLCNE									
Before	.26	.15	.26	.04	.30	.15	.27	.28	.27
3 months	.18	.10	.10	.13	.20	.18	.31	.30	.01
3 years	—	—	—	—	—	—	—	—	—
RLCSCT									
Before	.20	.10	.23	.05	.24	.12	.20	.21	.27
3 months	.01	−.04	−.08	.02	.00	.10	.12	.11	−.08
3 years	—	—	—	—	—	—	—	—	—
RLCSUT									
Before	.07	−.06	.31	.03	.18	.30	.25	.17	.30
3 months	.13	−.12	.18	.20	.22	.19	.25	.24	−.04
3 years	—	—	—	—	—	—	—	—	—
Jenkins Total									
Before	.16	.27	.16	.32*	.21	.47*	−.14	.18	.20
3 months	—	—	—	—	—	—	—	—	—
3 years	.27	.31	.38	.47*	.44*	.61**	−.05	.35	.30

(continued)

Table C-1. (*Continued*)

	Future despair	Social alienation	Somatic anxiety	Allergic inclination	Gastrointestinal susceptibility	Cardiovascular tendency	Pain proneness	Life threat	Emotional vulnerability
Months with Angina									
Before	.23	.18	.20	.23	.27	.05	−.06	.24	.04
3 months	—	—	—	—	—	—	—	—	—
3 years	—	—	—	—	—	—	—	—	—
Systolic Blood Pressure									
Before	.28	.27	.26	.41**	.29	.38*	.12	.24	.35*
3 months	—	—	—	—	—	—	—	—	—
3 years	—	—	—	—	—	—	—	—	—
Diastolic Blood Pressure									
Before	.20	.18	.16	.33*	.20	.39*	−.01	.25	.32*
3 months	—	—	—	—	—	—	—	—	—
3 years	—	—	—	—	—	—	—	—	—
Ejection Fraction prior to Surgery									
Before	.06	−.05	−.24	−.19	−.22	−.24	−.07	−.19	−.10
3 months	—	—	—	—	—	—	—	—	—
3 years	—	—	—	—	—	—	—	—	—

*$p < .05$ (two-tailed). **$p < .01$ (two-tailed). ***$p < .001$ (two-tailed).

8

Depression, Crisis Intervention, and Coronary Bypass Surgery

JUNE B. PIMM

DEPRESSION IN HEART ATTACK VICTIMS

Previous chapters have commented on the amount of depression experienced by many patients who have coronary bypass surgery. These have been clinical observations and serve to persuade us that depression exists with this surgery. However, these observations will not tell us much about who is most likely to be affected by depression or whether depression can be avoided or lessened by crisis intervention. This chapter will describe a research project that attempted to find out how frequently depression occurs in coronary bypass patients, who is most likely to suffer from this depression, and whether it can be helped by a crisis intervention program.

We can learn something about the frequency of depression from those who work with large numbers of patients who suffer heart attacks. Only half of our patients had experienced a myocardial infarction but all shared things in common with heart attack victims in that they suffered from debilitating angina. Angina is a constant warning that the heart is not functioning properly. Persons living with this condition exist in a state of vigilance which can create significant psychological stress.

For example, Katon (1982) states:

> In coronary artery disease, of 400,000 survivors of heart attack, 60% are estimated to display significant depression and anxiety during hospitalization. Twenty to thirty percent are found to be depressed one year later and

JUNE B. PIMM • Pimm Consultants, 2699 S. Bayshore Drive, Miami, Florida, 33133.

> studies have shown that fifteen to twenty percent of heart attack victims fail to return to work and that psychosocial factors play a major role.

Others who have been interested in psychiatric problems associated with illness, such as Crisp and his colleagues in London (Crisp, DeSouza, & Queenan, 1981), describe the psychological status of heart attack patients in this way:

> As a population, post infarct survivors are importantly more anxious and depressed, significantly more obsessional in their behavior and significantly more socially phobic and withdrawn than the normal population. They also complain a great deal more of somatic symptoms, some doubtless related to the consequences of their myocardial infarction. Such findings of a psychological kind remind us of the psychiatric morbidity and psychological crippling that can understandably follow the confrontation with death that arises with myocardial infarction, which may sometimes be amenable to wise counseling. (p. 6)

In contrast to professionals who describe general characteristics of large populations of patients, we also have the observations of those who have worked intensively with individual patients and who have written of their observations. Speedling followed eight patients and their families through the experience of a heart attack and described his findings in his book *Heart Attack* (1982). He observed,

> The act of healing cannot be complete until the social and emotional bonds which illness disrupted are themselves revitalized. As we have seen, psychosocial injury as consequences of physical injury can be acute. (p. 163)

Finally, the observations of a journalist whose husband, a physician, suffered a heart attack and later had coronary bypass surgery, formed the basis for the best seller *Heart Sounds* (Lear, 1980). Her observations had an impact on physicians and laymen alike.

Although we would generally agree that psychological distress might accompany or follow a heart attack, we might be surprised to know that depression can also *precede* this event. Evidence for this is provided by several researchers in the following studies. Crisp and associates described a longitudinal survey of both medical and psychological characteristics of a large sample of English patients. The research, done in South West London, was based on the records of all of the patients between the ages of 40 and 65 years who were registered with a group general practice. These patients were given a physical examination as well as a battery of psychological tests on three separate occasions (1969, 1971, and 1973). At the end of the 5-year study, an attempt was made to relate early psychological characteristics to later medical events.

Of the patients who later suffered myocardial infarctions, those who experienced heart attacks had significantly higher scores prior to their attack on the measure of somatic complaints (especially to do with sweating, heart pounding, and loss of libido). They were also significantly more depressed (answered positively to feelings of sadness). There appeared to be little relationship between later heart attacks and the reporting of physical symptoms to their doctors, nor did they consult their GPs more often than general.

Depression, as measured by admitting to "sad" feelings or by complaining of physical complaints, apparently can appear before a myocardial infarction. The researchers were interested in their observations that the heart attack patients endorsed both types of depressed responses—that is, sad feelings and physical complaints—more often than patients who did not later suffer an attack.

In the United States a longitudinal study was done by the Western Electric Company (Lebovits, Shekelle, Ostfeld, & Paul, 1967; Ostfeld, Lebovits, Shekelle, & Paul, 1964). Psychological information was obtained on 1,999 men who were reporting for their annual physical examination. These psychological evaluations were obtained on the first and fifth examination, and it was found that those who subsequently suffered heart attacks were significantly higher on the depression subscale of the MMPI than those who remained healthy. It was also noted that those who died from a myocardial infarction differed on their tests taken before the fatal heart attack from those who had a heart attack and survived.

In a somewhat different way of looking at depression, Dreyfull, Dasberg, and Assael (1969) looked at the incidence of myocardial infarction in hospitalized psychiatric patients. They discovered that myocardial infarction occurred almost exclusively in depressed patients. These authors suggested that, as the majority of the patients had developed their myocardial infarction long after the onset of their depression it was unlikely to be a reaction to the physical illness.

These writers and others who wrote several years ago (Pancheri, Bellaterra, Mateoli, Cristofari, Polizzi, & Pulette, 1978; Stern, Pascale, & Ackerman, 1977) all have found that depression and heart disease appear to be related. This fact presents a problem for surgeons who perform coronary bypass surgery. Concerns have been expressed regarding the amount of a patient's depression prior to surgery and the relationship of successful surgery to this depression (Kimball, 1969).

For example, researchers who looked at the psychosocial outcomes of coronary bypass surgery (Gundle Reeves, Tate, Raft, & McLaurin 1980) have not provided very optimistic news. They found that, despite good physical outcome, 83% of the patients were not employed after surgery, and 57% were sexually impaired 1–2 years later. Many of the patients were significantly limited in their life-style and had problems of low self-esteem, prolonged depression, and distorted body image. The authors conclude that more attention should be paid to the postsurgical psychological and social adjustment of bypass patients.

On the basis of these findings, as well as those that note such a high prevalence of depression in heart attack victims, we decided to try to ameliorate this depression through provision of crisis intervention. We will now describe some of the other studies which have attempted to provide helpful intervention to patients before we describe our own procedures and results.

INTERVENTION STUDIES

In finding crisis intervention an effective strategy in helping coronary bypass patients we add to the long list of studies in the literature. Many of these have produced similar results with patients recovering from heart attacks as well as patients who have undergone surgery.

Recent research findings suggest a strong need for follow-up programs for victims of heart attacks; they also suggest that the psychological and social characteristics of the patient will be important in the effectiveness of the rehabilitative effort (Cromwell, Butterfield, Brayfield, & Curry, 1977; Davidson, Winchester, Tayloe, Alderman, & Ingels, 1979; Fielding, 1979; Lenzner, 1974; Lenzner & Aronson, 1972; & Stocksmeier, 1976).

Segev and Schlesinger (1981) describe their interdisciplinary program for patients who have experienced a myocardial infarction at the Heart Institute of Assaf Harofe Hospital in Israel. The goals of the program were to increase the patient's life expectancy through decreasing the chances for another MI and to help patients and their families achieve psychological readjustment to their situation even if this entailed modification of their lifestyles. The program included three aspects: (1) medical treatment and physical training, (2) group discussion meetings with patients, and (3) group discussion meetings with patients' spouses. It was the conviction of those responsible for the program after eight years of experience that the rehabilitation program was responsible for remarkable changes in the patients' life-styles, improvements of family atmosphere, and adjustments of family members to the new situation.

Mayou (1981) also reports his experiences with offering various types of programs to cardiac patients. However, he suggests that three months after the patients' heart attacks there were only small benefits associated with programs. He found that most of the patients made excellent recovery from heart attack and argues that there is a strong argument for having intensive rehabilitation concentrate only on patients at risk for complications. His results suggest that "analysis of the pooled psychological and social data suggests criteria for identification of vulnerable patients." He would suggest confining specialized rehabilitation programs to these groups.

Naismith, Robinson, Shaw, and MacIntyre, (1979) studied the effectiveness of an intensive rehabilitation program on 143 men who had recently had a myocardial infarction. Half of the men were randomly allocated either to a group receiving intensive rehabilitation or to a control group. Their outcome was evaluated at the end of six months. The investigators found that patients who had neurotic, introverted personalities had a poor outcome in the control group but had a satisfactory outcome if they participated in the rehabilitation program. Neurotic personalities responded to help, and the rehabilitative measures did not increase their neurosis. The authors suggest that selection of patients who would benefit from psychological help seemed desirable.

These studies were interested in intervention with heart attack patients, but research has also demonstrated the effectiveness of psychological preparation for surgery. Unfortunately, however, there again the results have been mixed.

The classic study by Egbert, Battit, Welch, and Bartlett (1964) set the stage for optimism regarding the effectiveness of anticipatory guidance for surgical patients. In a study of 97 surgical patients, some were given specific information on what to expect during the postoperative period. They were also instructed on how to relax by deep breathing and how to move so that they would experience less discomfort from their stitches. The results showed that these patients were discharged from the hospital over two days earlier than the control patients. They also received less medication for pain than those who did not receive the presurgical information. Egbert's study has provided the model for most later research into the effect of presurgical information.

Prior to Egbert's study, Janis (1958) had written of the importance of matching "reassurance mechanisms" to the personality styles of patients experiencing surgery. For example, Janis found that for patients who express little fear and who appear to deny concern about the approaching surgery, the introduction of fear-arousing statements will induce the patient to begin the "work of worrying." For patients who on the other hand are showing an excessive amount of fear preoperatively, he suggests preparation which is aimed at reducing anxiety and calming the patient rather than providing information.

Later studies have found mixed effectiveness for presurgical interventions. Johnson (1981) reviews several of these studies, grouping them by type of interventions offered and type of result. For example, in five studies instruction in physical activities to perform after the operation were given to patients. All of these studies showed a reduction in negative emotional responses to surgery, but only two out of the five showed reduced length of hospitalization. She describes a study that utilized cognitive tasks, such as paying attention to the favorable aspect of the surgical experience, which resulted in reduced negative emotional responses and less pain medication but not less time in hospital. On the other hand, she reports research utilizing systematic relaxation plus sensory information that resulted in reduced length of hospitalization but not reduction in emotional responses. Concerning another study on the effect of individualized attention from a nurse, Johnson says the authors were able to achieve reduced emotional responses but not reduced hospital stays. Johnson reports on four other studies (three by her) that utilized information providing orientation to events accompanying surgery and their sequence of occurrence. All of these helped reduce emotional responsiveness, but only one showed an effect on length of hospital stay. In Johnson's opinion, the best results have been obtained by an approach which provides information focusing on the sensory components of the surgical experience. Two further studies conducted by Johnson demonstrated the effectiveness of this approach on both emotional responses and reduction in length of hospital stay.

Wood (1982) investigated the various spontaneous cognitive strategies used by 36 abdominal surgery patients during the immediate postoperative period. Personality factors were also studied. She decided that, although patients demonstrated a wide range of self-regulatory strategies, there was little to suggest that recovery quality or rate in this population was influenced by the cognitive strategies or personality variables.

Pickett and Clum (1982) looked at the relationship of locus of control and treatment strategies for the reduction of postsurgical pain and anxiety. They compared relaxation training, relaxation instructions, and an attention-redirection approach to no treatment in patients undergoing gall bladder surgery. Attention-redirection proved most effective in relieving postsurgical anxiety pain. The effectiveness of the cognitive distraction was greater for patients who were higher on the internal locus of control scale. This study is important in its utilization of a personality measure in addition to a treatment condition.

Wilson (1981) also utilized preoperative personality assessment of the variables denial, fear, and aggressiveness in research with elective surgery patients and concluded that behavioral preparation benefits even frightened aggressive or denying elective surgical patients.

Mumford, Schlesinger, and Glass (1982) reviewed 34 studies of psychological intervention with surgery and heart attack victims. They looked at 13 studies which used hospital days, after surgery or after heart attack as outcomes variables. On the average, psychological intervention reduced hospitalization approximately two days below the control group's average. In the 34 studies wherein patients were provided information or emotional support to help them master the crisis, they did better than patients who received only ordinary care. However, because some of these studies are also viewed negatively by MacDonald and Kuiper (1982), these conclusions should be viewed with caution. MacDonald and Kuiper were critical of studies which have investigated the psychosocial preparation for surgery. Their interest was in evaluating the effectiveness of these cognitive-behavioral approaches to reduction in the patient's pre- and postoperative anxiety with the ultimate goal of faster and less painful recovery. When preoperative levels of anxiety are taken into account, it appears that a more informative and supportive presurgical preparation might not be helpful for high- or low-anxiety patients. However, a cursory preparation for low-anxiety patients can be actually harmful. This is consistent with Janis's point of view. In looking at the different responses to treatment for copers versus avoiders, MacDonald and Kuiper suggest that the literature indicates that an information-based preparation may improve recovery for copers and hinder recovery for avoiders.

This research noted the frequency of the methodological flaws in many of the studies we have already reviewed. Flaws included lack of appropriate control groups to account for placebo and supportive counseling effects, or the lack of random assignment to groups, or differences in initial anxiety levels. The authors argue that few studies to date have been able to accurately demonstrate the effects of psychological preparation for pre- or postoperative anxiety.

Anderson and Masur (1983) reviewed studies concerned with preparation of patients, (both adults and children) for invasive medical and dental procedures. Their comparison included those which utilized a variety of preparatory approaches such as informative, psychotherapeutic, modeling, behavioral, cognitive-behavioral, and/or hypnotic. Like other reviewers, they conclude that serious methodological problems such as heterogeneity of sample characteristics, limited range of outcome measures, and lack of manipulation checks prevent definitive conclusion. They also express surprise, however, at the lack of interest in coming to a definite conclusion on this subject:

> Given that potential preparation benefits include the improved well being of the patient, reduced medical costs, and the fulfillment of legal-ethical requirements, it is surprising that health care providers are still debating whether such preparation is necessary.

INTERVENTION STUDIES WITH HEART SURGERY PATIENTS

There have been several studies specifically intended to help patients undergoing open-heart or coronary bypass surgery. The majority report results which are consistent with our findings. Aiken and Hendricks (1971), for example, compared the effects of a treatment group with a control group of patients receiving routine care. Treatment consisted of approximately 15 sessions of taped progressive relaxation and one session of supportive counseling by a nurse specialist. Fifteen male open-heart surgery patients showed significantly fewer psychiatric complications relative to the control group. MMPI scores showed a high incidence of depression in these patients, and the authors feel this is further evidence for the necessity of intervention with this type of surgery.

A study that compared the effects of a variety of in-hospital education approaches with coronary bypass patients (Barbarowicz, Nelson, DeBusk, & Haskell, 1980) used slide–sound education and compared it with the usual educational approach offered by the hospital. They found that increases in knowledge scores were significant for both teaching groups but that the scores for the slide–sound group remained significantly higher for three months. They also found that anxiety decreased significantly and that health-enhancing behavior increased significantly in both groups.

Owens and Hutelmyer (1982) looked at the effect of preoperative intervention on delirium in cardiac surgical patients. Sixty-four patients were divided into experimental or control groups, and members of the experimental group were informed in a preoperative interview of the possibility of experiencing temporary cognitive or sensory disturbances. They were reassured that these experiences were normal and encouraged to report them to the hospital staff. Postoperative interviews conducted from the fourth to eighth day postoperatively showed no statistically significant differences between experimental and control group. Many of them reported having unusual sensory or cognitive experiences. When the groups were compared regarding the person's feelings of comfort or control when confronted with an unusual experience, the difference between the experimental and control groups was significant.

Finally, a recent research project, *Recovery from Heart Surgery: Bio-Behavioral Factors,* at Boston University School of Medicine and Harvard Medical School looked at the perceived adequacy of patient education and fears and adjustments after cardiac surgery. The authors (Stanton, Jenkins, Savageau, Harkin, and Aucoin, in press) used a sample of 249 adults and followed them the first 6 months after surgery. They found that even patients who do perceive that they have been well prepared for the recovery process experience significant fears and adjustments after surgery. The authors feel that patients need additional reinforcement during the first 6 months after surgery to help alleviate these problems. All patients in the study received the hospital's prevailing methods of instruction, and 6 months later most felt that they had been well prepared as far as medical symptoms were concerned. More than half of the patients, however, felt that they had not been adequately prepared for possible emotional reactions and possible changes in the way other people would tend to treat them. Even more striking was the finding that patients with more fears and more adjustments to make following surgery also experienced worse angina at that time. The overall results of this study found that patients experienced fears and adjustments during the recovery period, even those who perceived that they had been well prepared. These were similar in number and intensity to those who reported that they were poorly prepared on an array of postoperative measures.

Finally, a study that did not succeed in helping patients with postsurgical delirium was reported by Surman, Hackett, Silverbert, and Behrendt (1974). This group of psychiatrists interviewed patients who were undergoing mitral valve surgery 48 hr before their operation. The purpose of the interview was to clear up any misconceptions the patient might have about the prospective surgery. They also taught the patients a simple autohypnotic technique. Their results showed no differences between the 20 patients in the informed group and 20 matched controls on such factors as amount of delirium, pain, anxiety, depression, or medication requirements. However, mitral valve surgery patients have been found to have different psychological reactions than do coronary bypass surgery patients (Willner, Rabiner, Wisoff, Hartstein, Struve, & Klein, 1976).

On the whole, the current literature supports the effectiveness of some kind of intervention program for patients undergoing surgery. However, problems of design and lack of adequate control groups often make it difficult to evaluate precisely the extent of the effect of intervention. Many writers suggest that there may be complicated interactions between such variables as personality, age, and severity of presurgical illness which contribute to the success of the intervention program as well (Bracken, Bracken, & Landry, 1977; Fitzgerald, McGowan, Kutner, & Wenger, 1982, Naismith *et al.*, 1979; Pickett & Clum, 1982).

CRISIS INTERVENTION AND HEART SURGERY PATIENTS

Our research design involved assessments of depression on all patients prior to surgery. We were interested in comparing these scores to depression scores after surgery for both crisis intervention and control patients. We found that our patients were mildly depressed on our measures before surgery, the average presurgical depression scores on the Beck Depression Inventory being approximately 9. While this is not clinically depressed in psychiatric terms, scores in this range are suggestive of some depression (Beck, 1978).

MEASUREMENT OF DEPRESSION

A word about our choice of depression measures might be helpful at this point: We needed to measure depression not only on the basis of clinical data such as case histories but also on an objective measure which could be subjected to statistical analyses. When we began to plan this study in 1975 the available scales which appeared most suitable were the Beck Depression Inventory and the MMPI Depression Scale. We also had subscale scores from the depression scale on the SCL-90-R.

The MMPI is viewed as a personality scale which describes a patient's basic personality traits and is used to facilitate understanding of his characteristic response pattern. Some studies using the MMPI as an instrument for assessing psychological change with coronary patients have found it unresponsive to the influence of short-term intervention (Martic, 1976). On the other hand, the Beck and SCL-90-R are thought to be instruments likely to pick up more transient mood states and thus capable of reflecting the effectiveness of therapy (Achtenberg & Lawlis, 1980). As it turned out, we randomly assigned patients to one or two groups, and by chance the two groups differed significantly on the MMPI pretest depression scale. Because their presurgical scores differed, their score differences after treatment could not be meaningfully evaluated and were dropped from the analysis.

We could have used the scores from the depression scale of the

SCL-90-R but were discouraged from doing so by the small number of items involved. Hoffman and Overall (1978) looked at the structure of the SCL-90-R scale and had some reservations about the use of factor score profiles with this measure. They suggest that

> the SCL-90-R tends to measure a unitary global complaint factor and that the self-report complaints of the patients are not highly differentiated with respect to the more specific factors the instrument may be capable of measuring.

This left us with the scores from the Beck Depression Inventory (Beck Ward, Mendelson, Mock, & Erbaugh, 1961). This scale has a lengthy and impressive literature, and because of its adequate test–retest reliability it appeared to be statistically the most appropriate instrument for our purposes. There have been a number of studies using the Beck with a variety of medical populations (Armstrong, Goldenberg, & Stewart, 1980; Coppen, Metcalfe, Carroll, & Morris, 1972; Fischer, 1967; Graff, 1965; Jain, 1971; Johnson & Heather, 1974; Khatami & Rush, 1978; Mayeux, Stern, Rosen, & Leventhal, 1981; Moffic & Paykel, 1975; Mohamed, Weisz, & Waring, 1978; Nielson & Williams, 1979; Schwab, Brown, & Holzer, 1968.)

Although these studies have looked at depression in a range of medical problems (Moffic & Paykel, 1975; Nielson & Williams, 1979) from obesity (Fischer, 1967; Graff, 1965) to Parkinson's disease (Coppen Metcalfe, Carroll, & Morris, 1972; Mayeux Stern, Rosen, & Leventhal 1981), we found only two that looked at depression in coronary patients (Fielding, 1979; Kolitz, 1983). As far as we know, there are none that investigate depression in coronary bypass patients.

THEORIES OF DEPRESSION

One of our problems was the theoretical underpinnings of the scale itself. The Beck scale was empirically derived and Beck states: "The items were chosen on the basis of their relationship to the overt behavioral manifestations of depression and do not reflect any theory regarding the etiology or the underlying psychological processes in depression" (Beck *et al.*, 1961, p. 562). However Beck's theory of depression is a cognitive theory (Beck, Rush, Shaw, & Emery, 1979), as he views depression as a negative view of events rather than a result of negative events. Therefore included in his scale are examples of items such as "I feel bad or unworthy a good part of the time" and "I feel I am being punished." These items are intended to reveal the extent to which a subject has adopted a pessimistic attitude toward life and is depressed because he views events in a negative way.

Because crisis intervention attempts to restore an individual's capacity to cope, it is more consistent with the theory of Seligman (1972, 1975), who has suggested that depression stems from "learned helplessness." His theory suggests that individuals who perceive themselves as unable to control their environments are susceptible to depression. Crisis intervention, which seeks to assist the individual recover adaptive coping strategies would appear to be more consistent with this model of depression.

Fortunately, Blaney (1977) in his critique of contemporary theories of depression, points out that the overlap between Beck's and Seligman's positions is considerable. Some of the cognitions to which Beck refers appear to be ones which might be expected of an individual in a state of helplessness. This would suggest that the Beck Depression scale would accurately reflect changes in depressed attitude after help in coping with coronary bypass surgery. Actually Seligman himself (Miller & Seligman, 1975) utilizes the Beck Depression Scale in his research with college students and depression.

Therefore Beck scores were obtained on all patients who entered our study and were completed in hospital the day before surgery. Patients were randomly assigned to groups and not matched on the basis of these scores. Unfortunately, because of this random selection, the patients who did not receive crisis intervention (control group) by chance had slightly higher Beck Depression scores (9.4) than patients who were randomly assigned to the crisis intervention treatment group (8.8). This difference, fortunately, was small and not statistically significant. Therefore the scores were subjected to statistical analysis.

Our results show that crisis intervention appears to have been effective in lowering depression in those patients who participated in this counseling. For the 34 patients whom we followed for three years, the Beck scores the day before surgery were 8.8, and three years after surgery the scores of these patients were 7.2. This strongly suggests that crisis intervention is useful with coronary bypass patients as a way of lowering depression after surgery. The patients who did not receive treatment, on the other hand, did not exhibit any significant change in their Beck Depression scores. The day before surgery their Beck scores were 9.4 and after three years they were still 9.1 (Figure 1).

These findings do not reach statistical significance but their direction provides support for the theoretical basis of crisis intervention. This theory suggests that upsetting life events can create an emotional crisis at the time they occur. This emotional reaction is reflected in our patients

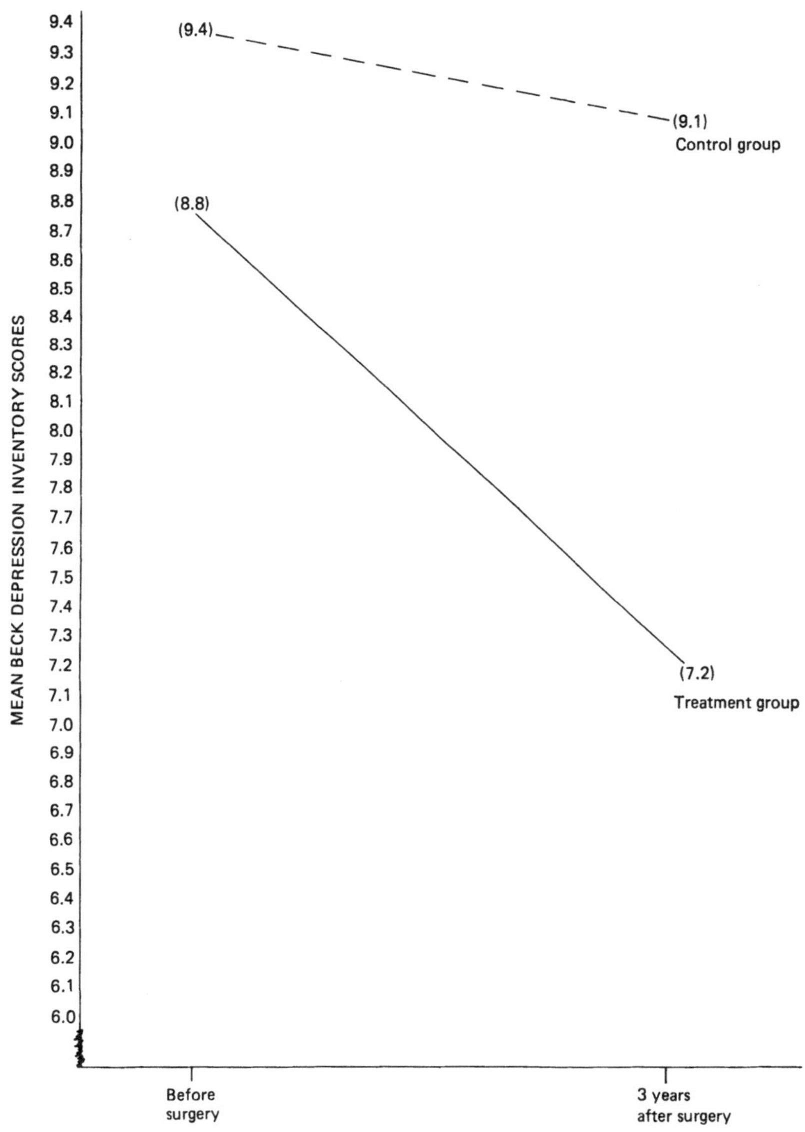

Figure 1. A comparison of Beck Depression Inventory scores before surgery and 3 years after surgery for 34 patients.

by slightly elevated depression scores immediately prior to surgery. Short-term supportive counseling resulted in depression scores within normal range for patients recovering from their surgical crisis. Patients who had no help in getting through the crisis, on the other hand, remained depressed for as much as three years after the successful surgical outcome.

This argument is based entirely on average scores and does not tell us much about the individual patients for whom this was true. If we categorize patients according to whether they are clinically depressed and compare those who are depressed with those who are not, the same picture emerges. Deciding upon a cutoff point for clinical depression is difficult. In looking at the literature with medical populations the cutoff score for clinical depression ranges from 9 to 14 (Baumgart & Oliver, 1981; Golin & Hartz, 1979; Meites, Lovallo, & Pishkin, 1980; Moffic & Paykel, 1975). On the other hand, some research uses average scores as we have done. If we are to utilize the score of 14 or over as signifying clinically depressed and compare our 34 patients on whom we have 3-year follow-up data, we get results as appear in Table 1.

We can look at the Beck scores in yet another way. Three years after surgery total scores on the Beck Depression Inventory were lower for the patients who received crisis intervention than they were the day before surgery. Therefore, they should not correlate highly with one another. This turned out to be the case, the correlation between presurgical Beck scores and three-year scores on treatment patients was only .464. For the control patients who did not receive crisis intervention, the correlation between presurgical and three-year Beck scores should be higher. This correlation was .832. The difference between these correlations was statistically significant and provides further evidence that crisis intervention was effective in helping patients cope with postsurgical depression.

DENIAL AND DEPRESSION

Finally, when we looked at individual items on the Beck, it appeared evident that only a portion of the items embedded in this measure could

Table 1. Percentage of Patients Having Scores of 14 or over on the Beck Depression Scale

	Before test	3 months after	3 years after
Treatment	19	14	6
Control	22	24	28

be termed cognitive in content. Some of them should be more appropriately viewed as affective, and a large proportion of them are clearly somatic. There is always concern about the validity of using bodily complaints as indices of depression in physically ill patients. This was pointed out by Crisp and his colleagues in the research reported earlier, wherein patients who suffered heart attacks presented the highest combination of depressive and somatic complaints. These authors argue for the "reasonableness" of these somatic responses in light of the patients' medical histories. On the other hand, bodily expressions or depression could reflect denial of emotional symptoms.

We became interested in the qualitative responses of our patients on the Beck Depression scale, especially in relation to the distinction between items which described cognitive-affective depression compared to those which were more closely identified with somatic complaints. It has been suggested that coronary bypass patients are more likely to use denial and express depression through somatic symptoms than patients who opt for medical treatment of their angina.

For example, since there is considerable controversy over medical or surgical treatment as the better method of treating angina (Hochman, 1982; Kolata, 1981), patients who prefer the "one shot" surgical approach might also have personalities different from those who choose the lifelong process of medical management. Williams, Haney, McKinnis, Harrell, Lee, Pryor, Califf, Kong, Rosati, and Blumenthal at Duke University (1980) suggest that over and above medical reasons for choosing one over the other, physicians should select surgery as appropriate treatment by prospectively identifying patients who are psychologically unlikely to respond to standard medical treatment of angina.

It is not unreasonable to assume that patients who prefer coronary bypass surgery to medical management are those more likely to deny their illness than those who are prepared to live with their condition. Medical management involves altering one's life-style in order to cope with disease, and this is incompatible with denial behavior. Components of medical management such as diet, exercise, taking of medication, and cessation of smoking are daily reminders of illness.

An example of denial seemed to us to be the extent to which our patients tended to deny the presence of angina as a reason for surgery. Interestingly, we found that 99% of our patients when asked, "Why did you have coronary bypass surgery?" responded that their doctors had told them it was necessary in order to save their lives. As Larson has pointed out (Chapter 2) surgeons inform patients that there are equally effective alternatives to surgery. Additionally, although patients were told that the purpose of surgery was the relief of angina, not necessarily prolongation of life, they obviously tended to distort this information.

Our findings are similar to those of McNeil and her colleagues (McNeil, Weichselbaum, & Pauker, 1978; McNeil, Weichselbaum, & Pauker, 1981; McNeil, Pauker, Sox, & Tversky, 1982), who present a fascinating discussion of the ways in which patients make decisions concerning surgery and how these decisions reflect their attitudes toward quality versus quantity of life.

Further evidence of the denial characteristics of our patients came from their responses to the question, "How long did you suffer from angina prior to surgery?" Only 73% of the patients reported experiencing angina. Of those who did experience angina pain, 90% reported that they had only experienced it for 6 months or less and 34% said they had experienced it for 1 month or less. On the other hand, information from the patients' records indicated that 96.8% of all patients had a positive history of angina. These patients had suffered angina for an average of 4½ years.

Anecdotally, our counselor reported that the majority of patients who agreed to participate in the study agreed to do so "in order to help science." They denied feelings of sadness or depression which might be ameliorated by counseling sessions. Some of our patients continued to attempt to run their offices from their beds in the hospital, implying that their lives were not disrupted to any degree by undergoing coronary bypass surgery.

Denial is not necessarily a negative coping strategy. Hackett and Cassem (1978) were the first to comment on the role of denial in reducing mortality and morbidity in coronary heart disease. They suggest that patients who are able to deny their worries have a better chance of surviving hospitalization for myocardial infarction. They also note that the way in which a patient will respond to myocardial infarction will be similar to the way he has dealt with stress in the past. In other words, typical coping styles can be predictive of a patient's capacity to cope with myocardial infarction.

Froese, Hackett, Cassem, and Silverberg (1974) looked at the relationship of denial to both anxiety and depression in patients in the coronary care unit. Their results suggest that there is a great deal of consistency and stability of denial across time. For example, it was rare to find those who deny shifting to a nondenying role. When occasionally this type of response did break down, it appeared to be a resilient defense that was quickly reinstated. These authors looked at the past histories of deniers and found that they usually responded to most of life's stresses and adversities by negating or minimizing threat. This pattern could be traced to young adulthood or adolescence. Their results also indicated that denial was not as effective in preventing depression in cardiac patients as it was in alleviating anxiety.

Probably Janis (1958) had the most influence on thinking about the effects of preoperative mood on postsurgical outcome. In his book *Psychological Stress: Psychoanalytic and Behavioral Studies of Surgical Patients* he describes three types of patients and their psychological reaction to surgery. Janis proposed that patients who had an extremely high level of anticipatory fear prior to surgery would be more likely than others to become fearful and upset after the experience. On the other hand, those who displayed an extremely low degree of anticipatory fear would also display psychological upset. These patients would be more likely to manifest their psychological problems through displays of anger or resentment.

The "low fear" subjects in Janis's studies were similar to patients whom others have labeled deniers. It appeared to be the case that denial of worry prior to surgery was an effective coping strategy if the surgical stress was minor but was not an adequate coping strategy for major surgical intervention. Janis concluded that patients who displayed a moderate amount of anticipatory fear before experiencing surgery appeared to have the best outcome. He described these patients as the ones who were capable of doing the "work of worrying" and suggested that a moderate amount of anticipatory worrying was necessary for adaptation to severe stress.

If our patients had a tendency to deny, we reasoned that this would be reflected in their specific responses on the Beck before and after their surgery. For example, we expected that they would initially endorse at least as many somatic or bodily responses as cognitive or affective. This expectation was based on the fact that they were physically ill and had been living with severe angina for a significant period before we interacted with them. For deniers, somatic symptoms could provide a socially acceptable way of dealing with feelings.

Accordingly, we divided the items on the Beck into two groups similar to the way in which Leiber, Plumb, Gerstenzang, and Holland (1976) looked at cancer patients. One group of responses seemed to tap cognitive and affective responses whereas another group seemed to deal more with somatic depressive symptoms.

After surgery, we expected treatment patients to endorse fewer somatic responses. This expectation was based on two possibilities. Either patients who denied depression before surgery (expressing it instead through the socially acceptable bodily responses) were no longer depressed or the bodily complaints had been genuine expressions of discomfort. In other words, after successful medical recovery from their coronary bypass surgery these real symptoms would no longer be present and therefore fewer would be endorsed. Conversely, if the crisis of coronary bypass surgery creates a psychological state of disequilibrium which if not adequately resolved leaves a patient with depression, we would expect our control patients to continue to endorse more items than treatment patients.

As we had expected, our treatment patients had fewer somatic complaints after surgery, and cognitive-affective responses remain at about the same level (Figure 2). This is consistent with crisis intervention theory, which does not purport to alter precrisis personality styles but hopes to help alleviate maladaptations precipitated by a crisis. On the other hand, our control, or no-treatment patients, continued to endorse approximately the same number of somatic complaints (Figure 3). Since this occurred in spite of satisfactory medical and surgical outcome, it suggests that endorsement of bodily complaints is more likely evidence of depression than expression of physical discomfort. Because both groups had similar physical outcomes after surgery, treatment and control groups should not have differed in the number of somatic items

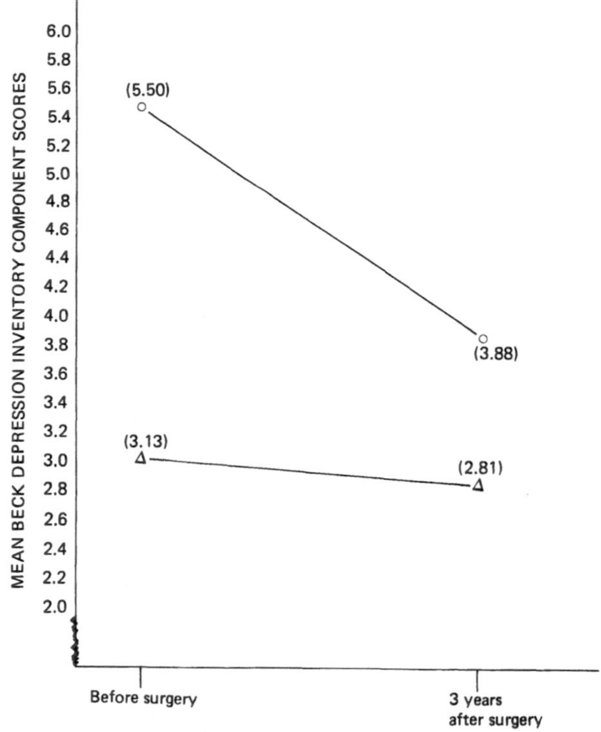

Figure 2. A comparison of cognitive and affective (△) versus somatic (○) items on the Beck Depression Inventory before surgery and three years after discharge for treatment patients.

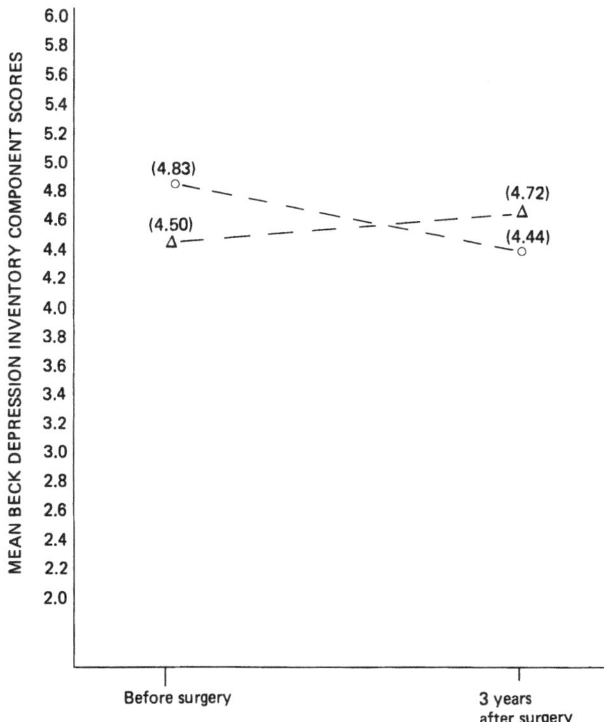

Figure 3. A comparison of cognitive and affective (△) versus somatic (○) items on the Beck Depression Inventory before surgery and 3 years after discharge for control patients.

endorsed. These results suggest that patients who did not receive crisis intervention may not feel they benefitted as much from the surgery even with a good physical outcome.

BEHAVIORAL DEFINITIONS OF DEPRESSION

Another example of the possible denial of overt depression in our patients comes from the results of the SCL-90-R Analogue. Using this device, the wives of the patients were asked to rate their husbands' depression after surgery. This required an assessment of the extent of the patient's depression based on her observations of his behavior.

Behavioral observations are considered good indicators of depression (Katon, 1982; Weiss, Bailey, Goodman, Hoffman, Ambrose, Salman, & Charry 1982). For example, Katon states, "In most of non-Western, as well as much of Western cultures, major depression is experienced as a somatic disorder with denial and minimization of affective and cognitive symptoms." In his opinion, for the diagnosis of depression the following behavioral characteristics must be present nearly every day for a period of at least one week to one month:

1. Poor appetite or significant weight loss or increased appetite and significant weight gain
2. Insomnia or hypersomnia
3. Psychomotor agitation or retardation
4. Loss of interest or pleasure in usual activities or decreased sexual drive
5. Loss of energy or fatigue

Animal research also has to rely on behavioral indices when investigating depression. The work of Weiss *et al.* (1982) is a particularly important example. Weiss has been able to elicit depression in rats by exposing them to electric shock and allowing them no opportunity to escape. This produces behaviors which are similar to those demonstrated by humans who are diagnosed as clinically depressed. Weiss relates these behavioral manifestations of depression to the presence of, and changes in, level of norepinephrine in the area of the brain called the locus coeruleus. Weiss argues that this research provides definitive evidence of the relationship between experience (loss of ability to cope, or uncontrollability of events) and the appearance of depression. He has also identified the underlying mechanism for the depressed state as neurochemical activity in the locus coeruleus.

In our research, using the SCL-90 R Analogue, 3 years after surgery, patients who received crisis intervention were rated as significantly less depressed (11.1) than those who did not (30.5). This finding was statistically significant and surprisingly did not appear until the 3-year measure. When assessing spouses at 12 weeks after surgery, the wives of patients did not report the same amount of depression in their husbands (see Chapter 7, Figure 1).

Also, to look at the same data another way, for crisis intervention patients the correlation between the Analogue scores at 12 weeks and 3 years was high (.819), as these scores were low on both occasions, indicative of little or no depression. However, because the Analogue Depression scores for patients who did not receive crisis intervention began low but increased substantially from 12 weeks after surgery to 3 years after surgery, their correlation is low (.273). The difference between these correlations is statistically significant.

Twelve weeks after surgery, the wives of control patients might have assumed that too little time had elapsed since surgery for depressed behavior to disappear. Therefore they did not indicate depression for their husbands on the SCL-90-R Analogue. Three years later, on the other hand, wives of the same patients were much more willing to admit to the depressed behavior they could observe in their husbands, and they answered differently on the Analogue.

In summary, we can say on the basis of the Beck scores and the responses on the SCL-90-R Analogue that patients who participated in crisis intervention were less depressed than patients who did not. Initially, the responses of the patients on the Beck Depression Inventory were moderately high and their pattern of responses was suggestive of a tendency to deny. However, patients who received short-term counseling appeared to have less need to respond with somatic manifestations of depression and presented a more normal picture on the Beck Depression Inventory.

The wives of patients, on the other hand, seemed reluctant to recognize depressed behavior in their husbands 12 weeks after surgery. This was true even when patients themselves were endorsing a moderately high number of items on the Beck. At the end of 3 years, however, the same wives endorsed a significant level of depression in their husbands. We have speculated that wives may have needed time to accept the depressive behavior in their husbands and that 12 weeks was too close to surgery for them to be concerned.

Our research found depression in coronary bypass patients and demonstrated that those who participated in crisis intervention were less depressed. However, this was obviously not true of all patients in the study. The next chapter will describe individual differences in our patient population and the ways in which these differences in personality characteristics interacted with counseling and its effect on depression. To conclude this chapter however, we present two case histories. One is a patient who appeared to benefit significantly from crisis intervention in terms of a significantly lowered Beck score; the other is a patient in the control group who became much more depressed after surgery.

CASE HISTORIES

Patient A (Control Group)

	Before test	12 weeks	3 years
SCL-90-R GSI	57	53	57
Beck	5	9	10
Locus of Control	3	NA	7
Analogue of Depression	—	20	80
Recent Life Changes	2	3	—
Type A			

This case is a 66-year-old man who had been semi-retired before surgery. He was married, had two children still at home, and had been living in the community for 26 years. He had been suffering with angina for approximately 10 years; when it had been more severe for 5 months surgery was suggested. His medical record shows that he had triple coronary disease and an ejection fraction of 71% and was a New York Heart Association Class 3. He had left main coronary disease and five bypasses at surgery.

On his presurgical tests this patient was a Type A on the Jenkins Activity Scale. He was not reporting a significant number of symptoms on the SCL-90-R. He was internal in his locus of control and did not appear significantly depressed on the Beck Depression Scale.

Twelve weeks after surgery this patient again did not report a significant number of symptoms on the SCL-90-R and did not respond to the locus of control scale; he was reporting more depression on the Beck Depression Inventory, but his wife was not reporting this depression on the SCL-90-R Analogue. At 3 years, although his SCL-90-R score was still considered normal, he was reporting significant depression on the Beck and his wife was not recognizing his depression on the SCL-90-R Analogue. Neither before surgery nor 12 weeks after surgery did this patient indicate a large number of recent life changes.

This patient's medical recovery was marred by complications and he had to be hospitalized again after surgery for coughing and difficulty breathing. Because he was assigned to the control group, there was no contact with him between surgery and the 12-week follow-up, at which time he reported dissatisfaction with the surgical outcome. When asked if he were to have to repeat the surgery, he replied that he did not think he would. His wife reported that both she and her husband had felt very inadequately prepared for his surgical and in-hospital experience.

All patients in the 3-year follow-up group were interviewed by a counselor new to the project who did not have access to previous records. Three years later, the patient did not appear to be doing well physically or psychologically. He was taking several medications and still complained of symptoms such as coughing, angina, and inability to think quickly. He felt an overwhelming sense of fatigue and reported that prior to surgery he had been moderately active, whereas he now described himself as sedentary. During the interview the patient constantly reiterated his frustration at his poor memory and loss of intellectual efficiency and reported that the results of the surgery were "not as good as he had hoped for." This patient's relationship with his doctor was poor as demonstrated by the patient's rating of the doctor as unsatisfactory in terms of availability and general satisfaction; under the section "provision of services" instead of a rating the patient put in a dollar sign!

The patient reported that he now participated less in social activities, hobbies, clubs, and organizations than he did before surgery. His wife confided to the interviewer that her husband appeared changed since his surgery and that she would have welcomed counseling during the recovery process if it had been made available to her.

This case appears to have been one in which a coronary bypass surgery did not improve the patient's quality of life. Not only does he complain of psychological problems but he is less active and considers himself physically handicapped. This case is in contrast to that of Patient B.

Patient B (Treatment Group)

	Before test	12 weeks	3 years
SCL-90-R GSI	62	57	51
Beck	11	6	2
Analogue of Depression	—	3	0
Locus of Control	NA	0	1
Recent Life Changes	—	—	—
Type B			

This is a 70-year-old man, married, and with one child living in the community. He has lived there for 36 years and had been self-employed prior to surgery. At the time of his operation he was distressed over some business reverses.

Medically, Patient B is very similar to Patient A: a history of angina for approximately 5 months prior to surgery, left main artery disease, triple coronary disease, and five bypasses. Prior to surgery, his ejection fraction was 64% and he was New York Heart Association Class 3.

This patient was assigned to the crisis intervention group and the counselor recalls him being a very difficult case. The patient was married a second time and there were unresolved difficulties in his relationship with his wife, including sexual problems. Neither the patient nor his wife was comfortable in their roles with one another and the crisis of surgery exacerbated these problems. The patient was anxious to be independent and felt that his wife was overprotective; on the other hand, when she responded to this by going out and getting a job, he felt threatened.

This patient also had a difficult recovery in the hospital. He became anxious and depressed when moved out of the intensive care unit and needed much support and help from the counselor both in the hospital and when he was discharged to his home. Weekly visits of counseling were scheduled, and in addition there were many phone calls from both the patient and his wife between visits. Usually the purpose of the call was to ask for help in dealing with the depression or anxiety of the spouse, regardless of which one made the call.

Prior to surgery, this patient was classified as a Type B; he had an SCL-90-R score within normal limits and a Beck score of 11, which signifies depression. He did not respond to the locus of control scale. At 12 weeks his SCL-90-R score was still normal, but his Beck score had dropped to 6 and his wife was not reporting him as depressed on the SCL-90 Analogue. At 3 years he had a Beck score of 3, which is clearly not depressed, and his wife again reported no depression. The patient did not respond to the Recent Life Changes Scale and had a score of 1 on locus of control which suggests internal locus of control.

The counselor used a crisis intervention approach throughout, restricting her help to the crisis at hand rather than attempting to deal with the underlying marital problems. Three years later, this couple is doing well. At the time of the follow-up interview the patient reported that he felt the results of his surgery were better than expected. Although his working day has been reduced, he is more actively engaged in social activities, sports, hobbies, and clubs. He particularly enjoys his participation in the Heart to Heart Club, and his wife now works as a volunteer at the hospital.

The patient also appears to have made an excellent physical recovery. Initially he made frequent telephone calls to his doctor with concern about physical symptoms. Now he does not need any medication, and states that he watches his diet and exercises regularly. Both the patient and his wife report that they feel the counseling made all the difference between a successful or unsuccessful outcome.

REFERENCES

Achtenberg, J. A., & Lawlis, G. F. *Bridges of the bodymind.* Champaign, Illinois: Institute for Personality and Ability Testing, 1980.

Aiken, L. H., & Henrichs, T. F. Systematic relaxation as a nursing intervention technique with open heart surgery patients. *Nursing Research,* 1971, *20* (3), 212–216.

Anderson, K. O., & Masur, F. T., III. Psychological preparation for invasive medical and dental procedures. *Journal of Behavioral Medicine,* 1983, *6* (1), 1–40.

Armstrong, H. E., Goldenberg, E., & Stewart, D. Correlations between Beck

Depression scores and physical complaints. *Psychological Reports,* 1980, *46,* 740–742.

Barbarowicz, P., Nelson, M., DeBusk, R. F., & Haskell, W. L. A comparison of in-hospital education approaches for coronary bypass patients. *Heart and Lung,* 1980, *9* (1), 127–132.

Baumgart, E. P., & Oliver, J. M. Sex ratio and gender difference in depression in an unselected adult population. *Journal of Clinical Psychology,* 1981, *37,* 570–574.

Beck, A. T. *The Beck Depression Inventory.* Philadelphia: Philadelphia Center for Cognitive Therapy, 1978.

Beck, A. T., Rush, J. A., Shaw, B. F., & Emery, G. *Cognitive therapy of depression.* New York: Guildford, 1979.

Beck, A. T., Ward, C. H., Mendelson, M., Mock, J., & Erbaugh, J. An inventory for measuring depression. *Archives of General Psychiatry,* 1961, *4,* 561–571.

Blaney, P. M. Contemporary theories of depression: Critique and comparison. *Journal of Abnormal Psychology,* 1977, *86*(3), 203–223.

Bracken, M. B., Bracken, M., & Landry, A. B., Jr. Patient education by videotape after myocardial infarction: An empirical evaluation. *Archives of Physical Medicine and Rehabilitation,* 1977, *58,* 213–219.

Coppen, A., Metcalfe, M., Carroll, J. D., & Morris, J. G. L. Levodopa and L-tryptophan therapy in parkinsonism. *The Lancet,* 1972, *1,* 654–658.

Crisp, A. H., DeSouza, M., & Queenan, M. *Myocardial infarction and the emotional climate.* Paper presented at the Sixth World Congress of the International College of Psychosomatic Medicine, Montreal, Quebec, Canada, Sept. 13–18, 1981.

Cromwell, R. L., Butterfield, E. C., Brayfield, F. M., & Curry, J. J. *Acute myocardial infarction: Reaction recovery.* St. Louis: Mosby, 1977.

Davidson, D. M., Winchester, M. A., Tayloe, B. C., Alderman, E. A., & Ingels, N. B., Jr. Effects of relaxation therapy on cardiac heart disease. *Psychosomatic Medicine,* 1979, *41*(4), 303–309.

Dreyfuss, F., Dasberg, H., & Assael, M. F. The relationship of myocardial infarction to depressive illness. *Psychotherapy and Psychosomatics,* 1969, *266,* 796–801.

Egbert, L. D., Battit, G. E., Welch, C. E., & Bartlett, M. K. Reduction of postoperative pain by encouragement and instruction of patients. *New England Journal of Medicine,* 1964, *270,* 825–827.

Fielding, R. Behavioural treatment in the rehabilitation of myocardial infarction patients. In D. J. Osborne & M. M. Gruneberg (Eds.), *Research in psychology and medicine* (Vol. 1). New York: Academic Press, 1979.

Fischer, N. Obesity, affect, and therapeutic starvation. *Archives of General Psychiatry,* 1967, *17,* 227–233.

Fitzgerald, T., McGowan, D., Kutner, M., & Wenger, N. K. Demographic determinants of success in the vocational rehabilitation of cardiac patients. *Journal of Rehabilitation,* 1982, *48* (2), 35–38.

Froese, A., Hackett, T. P., Cassem, N. H., & Silverberg, E. L. Trajectories of anxiety and depression in denying and nondenying acute myocardial infarction patients during hospitalization. *Journal of Psychosomatic Research,* 1974, *18,* 413–420.

Golin, A., & Hartz, M. A. A factor analysis of the Beck Depression Inventory in a mildly depressed population. *Journal of Clinical Psychology,* 1979, *35,* 323–325.

Graff, H. Overweight and emotions in the obesity clinic. *Psychosomatics,* 1965, *6,* 89–94.

Gundle, M. J., Reeves, B. R., Jr., Tate, S., Raft, D., & McLaurin, L. P. Psychosocial outcome after coronary artery surgery. *American Journal of Psychiatry,* 1980, *137* (12), 1591–1594.

Hackett, T. P., & Cassem, H. H. Psychologic aspects of rehabilitation after myocardial infarction. In N. D. Wenger & H. K. Hellerstein (Eds.), *Rehabilitation of the coronary patient.* New York: Wiley, 1979.

Hochman, G. *Heart bypass—What every patient must know.* New York: St. Martin's Press, 1982.

Hoffmann, N. G., & Overall, P. B. Factor structure of the SCL-90 in a psychiatric population. *Journal of Consulting and Clinical Psychology,* 1978, *46,* 1187–1191.

Holm, K., & Carlsen, P. M. Is there a best approach to teaching cardiac patients? *American Journal of Nursing,* 1982, 287.

Jain, V. K. Affective disturbance in hypothyroidism. *British Journal of Psychiatry,* 1971, *119,* 279–280.

Janis, I. L. *Psychological stress: Psychoanalytic and behavioral studies of surgical patients.* New York: Wiley, 1958.

Johnson, D. A. W., & Heather, B. B. The sensitivity of the Beck Depression Inventory to changes of symptomatology. *British Journal of Psychiatry,* 1974, *125,* 184–185.

Johnson, J. D. *Interventions and their effect on emotional response and length of hospitalization.* Paper presented at the American Psychological Association Conference, Montreal, Canada, 1981.

Katon, W. Depression: Somatic symptoms and medical disorders in primary care. *Comprehensive Psychiatry,* 1982, *23* (3), 274–287.

Khatami, M., & Rush, A. J. A pilot study of the treatment of outpatients with chronic pain: Symptom control, stimulus control and social system intervention. *Pain,* 1978, *5,* 164–172.

Kimball, C. P. Psychological responses to the experience of open heart surgery. *American Journal of Psychiatry,* 1969, *126,* 348.

Kolata, G. B. Consensus on bypass surgery. *Science,* 1981, *211* (2), 42–43.

Kolitz, S. L. *The effect of personality style and depression upon the physical and emotional recuperation of patients who have sustained a first myocardial infarction.* Unpublished manuscript, Department of Psychology, University of Miami, Coral Gables, Florida, June 1983.

Lear, M. W. *Heart sounds.* New York: Simon & Schuster, 1980.

Lebovits, B. Z., Shekelle, R. B., Ostfeld, A. M., & Paul, O. Prospective and retrospective psychological studies of coronary heart disease. *Psychosomatic Medicine,* 1967, *29*(3), 265–272.

Leiber, L., Plumb, M. M., Gerstenzang, M. L., & Holland, J. The communication of affection between cancer patients and their spouses. *Psychosomatic Medicine,* 1976, *38,* 379–389.

Lenzner, A. S. Psychiatry in the coronary care unit. Reprinted from *Psychosomatics,* 1974, *14,* 70–71.

Lenzner, A. S., & Aronson, A. L. Psychiatric vignettes from a coronary care unit. Reprinted from *Psychosomatics,* 1972, *13*(3), 179–184.

MacDonald, M. R., & Kuiper, N. A. *Psychosocial preparation for surgery: Theoretical and methodological critique.* Paper presentation at the Annual Meeting of the Canadian Psychological Association, Montreal, Canada, 1981.

Mayeux, R., Stern, Y., Rosen, J., & Leventhal, J. Depression, intellectual impairment, and Parkinson disease. *Neurology,* 1981, *31,* 645–650.

Martic, M. Results of psychological testing of coronary paths in a longitudinal study of the following up of effects of training. In U. Stocksmeier (Ed.), *Psychological approach to the rehabilitation of coronary patients.* Berlin: Springer-Verlag, 1976.

Mayou, R. Effectiveness of cardiac rehabilitation. *Journal of Psychosomatic Research,* 1981, *25* (5), 423–427.

McNeil, B. J., Weichselbaum, R., & Pauker, S. G. Fallacy of the five-year survival in lung cancer. *New England Journal of Medicine,* 1978, *299,* 1397–1401.

McNeil, B. J., Weichselbaum, R., & Pauker, S. G. Speech and survival: Tradeoffs between quality and quantity of life in laryngeal cancer. *New England Journal of Medicine,* 1981, *305,* 982–987.

McNeil, B. J., Pauker, S. G., Sox, H. C., & Tversky, A. On the elicitation of preferences for alternative therapies. *New England Journal of Medicine,* 1982, *306,* 1259–1262.

Meites, K., Lovallo, W., & Pishkin, V. A comparison of four scales for anxiety, depression, and neuroticism. *Journal of Clinical Psychology,* 1980, *36,* 427–432.

Miller, W. R., & Seligman, M. E. P. Depression and learned helplessness in man. *Journal of Abnormal Psychology,* 1975, *84,* 228–238.

Moffic, H. S., & Paykel, E. S. Depression in medical in-patients. *British Journal of Psychiatry,* 1975, *126,* 346–353.

Mohamed, S. M., Weisz, G. M., & Waring, E. M. The relationship of chronic pain to depression, marital adjustment, and family dynamics. *Pain,* 1978 *5,* 285–292.

Mumford, E., Schlesinger, H. J., & Glass, G. V. The effects of psychological intervention on recovery from surgery and heart attacks: An analysis of the literature. *American Journal of Public Health,* 1982, 72 (2), 141–151.

Naismith, L. D., Robinson, J. F., Shaw, G. B., & MacIntrye, M. M. Papers and originals, *British Medical Journal,* 1979, Feb. 17, 439–441.

Neilsen, A. C., & Williams, T. A. Depression in ambulatory medical patients. *Archives of General Psychiatry,* 1980, *37,* 999–1004.

Ostfeld, A. M., Lebovits, B. Z., Shekelle, R. B., & Paul, O. A prospective study of the relationship between personality and coronary heart disease. *Journal of Chronic Diseases,* 1964, *17,* 265–276.

Owens, J. F., & Hutelmyer, C. M. The effect of preoperative intervention on delirium in cardiac surgical patients. *Nursing Research,* 1982, *31* (1), 60–62.

Pancheri, P., Bellaterra, M., Mateoli, S., Cristofari, M., Polizzi, C., & Pulette, M. Infarct as a stress agent: Life history and personality characteristics in improved versus not-improved patients after severe heart attack. *Journal of Human Stress,* 1978, March 16–22.

Pickett, C., & Clum, G. A. Comparative treatment strategies and their interaction with locus of control in the reduction of postsurgical pain and anxiety. *Journal of Consulting and Clinical Psychology,* 1982, *50* (3), 439–441.

Schwab, J. J., Brown, J. M., & Holzer, C. E. Depression in medical inpatients with gastrointestinal disease. *American Journal of Gastroenterology,* 1968, *49,* 146–152.

Segev, U., & Schlesinger, Z. Rehabilitation of patients after acute myocardial infarction–An interdisciplinary, family-oriented program. *Heart and Lung,* 1981, *10* (5), 841–847.

Seligman, M. E. P. Learned helplessness. *Annual Revue of Medicine,* 1972, *23,* 407–412.

Seligman, M. E. P. *Helplessness—On depression, development and death.* San Francisco: Freeman, 1975.

Speedling, E. J. *Heart attack: The family response at home and in the hospital.* New York: Tavistock Publications, 1982.

Stanton, B., Jenkins, C. D., Savageau, J. A., Harken, D. E., & Aucoin, R. Perceived adequacy of patient-education and fears and adjustments after cardiac surgery. *Heart and Lung,* in press.

Stern, M. J., Pascale, L., & Ackerman, A. Life adjustment postmyocardial information. *Archives of Internal Medicine,* 1977, *137,* 1680–1685.

Stocksmeier, U. *Psychological approach to the rehabilitation of coronary patients.* Berlin: Springer-Verlag, 1976.

Surman, O. S., Hackett, T. P., Silverberg, E. L., & Behrendt, D. M. Usefulness of psychiatric intervention in patients undergoing cardiac surgery. *Archives of General Psychiatry,* 1974, *30,* 830–835.

Weiss, J. M., Bailey, W. H., Goodman, P. A., Hoffman, L. J., Ambrose, M. J., Salman, S., & Charry, J. M. A model for neurochemical study of depression. In N. Y. Spiegelstein & A. Levy (Eds.), *Behavioral models and the analysis of drug action. Proceedings of the 27th OHOLO Conference,* Zichron Ya'acov, Israel 28–31, March, 1982. Amsterdam: Elsevier Scientific Publishing Company, 1982.

Williams, R. B., Jr., Haney, T. L., McKinnis, R. A., Harrell, F. E., Jr., Lee, K. L., Pryor, D. B., Califf, R., Kong, Y. H., Rosati, R. A., & Blumenthal, J. A. *Psychosocial and physical predictors of anginal pain relief with medical management.* Paper presented at the annual meeting of the American Heart Association, Miami, Florida, November 1980.

Willner, A. E., Rabiner, C. J., Wisoff, B. G., Hartstein, M., Struve, F. A., & Klein, D. F. Analogical reasoning and post operative outcome. *Archives of General Psychiatry,* 1976, *33,* 255–259.

Wood, D. J., & Pesut, D. J., Psychosomatic self-regulation and recovery from surgery. *Nursing Research,* 1982, *31* (3), 191.

9

The Psychology of Coronary Bypass Patients

JUNE B. PIMM

Patients who experience coronary bypass surgery are likely to suffer depression, and previous chapters have described an intervention strategy that appears to help. In this chapter I shall describe the personality characteristics of these patients, especially those characteristics which can predict those who will benefit most from crisis intervention.

All who have worked with coronary bypass patients and who have attempted intervention strategies have emphasized the importance of personality characteristics. They suggest that different personal characteristics will aid or impede recovery and also interact with programs aimed at helping patients cope with the surgical procedure (Blachly & Blachly, 1968; Brown & Rawlinson, 1975, 1976; Merwin & Abram, 1977; Pilowski, Spence, & Waddy, 1979; Ramshaw & Stanley, 1981; Thurer, Levine, & Thurer, 1980–81; Wilson-Barnett, 1981; Zyzanski, Stanton, Jenkins, & Klein, 1981).

Our findings suggest that the personality characteristics of our patients were typical of coronary patients in some ways; however, in other ways they appeared to differ from other groups of patients with heart problems. One of the surprising findings of our study was the extent to which medical characteristics were predictive of depression and the manner in which medical and psychological factors interacted with crisis intervention.

The following sections will describe our psychological measures and discuss the ways in which they interacted with depression. We will also

JUNE B. PIMM • Pimm Consultants, 2699 S. Bayshore Drive, Miami, Florida 33133.

describe a theoretical formula for predicting which surgical patients will most likely need crisis intervention. However, in order that the reader may appreciate the implications of our results, the chapter will begin with a brief review of the present state of the art in health psychology.

In general, the field of health psychology has emphasized the relationship of illness and stress. Selye (1974), the pioneer in the field of stress studies, emphasized the highly personalized aspect of stress and pointed out that stress is actually "in the eye of the beholder." For stress to cause distress, in Selye's terms, it should be termed a *stressor*. This point of view has been extended by those who have emphasized coping ability and its relationship to illness (Lazarus, 1974; Weiss, 1972). In other words, a stressful experience may not have harmful effects if the individual possesses adequate coping mechanisms for dealing with the event. Closely tied to this thinking has been the research on "learned helplessness" leading to depression (Seligman, 1972; Weiss, Bailey, Goodman, Hoffman, Ambrose, Salman, & Charry, 1982).

Our measures have attempted to clarify some of these issues by attempting to assess stress through the Recent Life Changes Questionnaire Scale. Coping strategies have been evaluated by the Locus of Control Scale and the Jenkins measure of coronary prone behavior (Type A). The results of each of these measures will be presented individually. Since these measures interact with each other in a number of ways and also with medical characteristics, the last part of the chapter will describe these relationships.

STRESS

Hans Selye is generally credited with the introduction of stress to the field of medicine. In his book *Stress without Distress* (1974), Selye coined the term *General Adaptation Syndrome* or GAS, which he used to describe the physiological changes which occur when an individual or animal is exposed to stress. The initial phase of the GAS consists of an alarm reaction during which there is an increase in physiological responding. This Selye was able to demonstrate in laboratory animals by exposing them to severe conditions of cold or muscular effort. He observed that if these animals were exposed to noxious conditions for too long a period of time they would die. At autopsy they would show characteristic changes in their adrenal glands, thymus glands, and lymph nodes. Selye suggested that these animals had marshalled physiological defenses in order to adapt to the noxious stimuli and that for a period, which he called the *stage of resistance*, the animals were capable of coping. After a prolonged period of resistance, however, inevitably the strain became too great and the animal entered the final phase of the GAS, which is the stage of exhaustion. At that point, death usually occurred because of severe damage to one of the body organs. Selye suggests that through accidental conditioning various parts of the body will become weakened and that the weakest part of the body (heart, kidney, gastrointestinal tract, or brain) will break down first under the stress. Figure 1 presents a diagram of Selye's model (Cohen, 1980).

Other researchers in the field of stress (Weiss, 1972) have been able to demonstrate in work with animals that psychological stress can result in the development of physical illness. For example, Weiss exposed two rats to the identical number of shocks but allowed one rat to predict when to expect the shock. The rat which received shock unpredictably developed a significant amount of ulceration of the stomach. Because Weiss felt that the psychological factor of unpredictability was the main determinant of ulcer severity, he then did a series of experiments in which he studied what he called coping behavior. He paired rats so that one rat could either escape shock or at least terminate it quickly once it had started, the other rat would receive whatever shocks its partner was unable to avoid. In other words, both rats received the same number of shocks, but one rat was in a position to attempt to avoid or escape shock entirely. In psychological terms one rat was "helpless" and the other rat had "control" or was able to "cope." As in the earlier research, there was a significant difference in the amount of stomach ulceration between the rat that was able to control the number of shocks and the animal that was unable to do so.

More recently Lazarus (1974) has applied these ideas to human illness by postulating the importance of cognition or appraisal of stress and its effect on ability to cope with stress. Lazarus theorizes that coping strategies provide the intervening variable between stress and disease and has amended Selye's original model for GAS. Figure 2 depicts this adaptation (Cohen, 1980; Stone, Cohen, & Adler, 1980, p. 83).

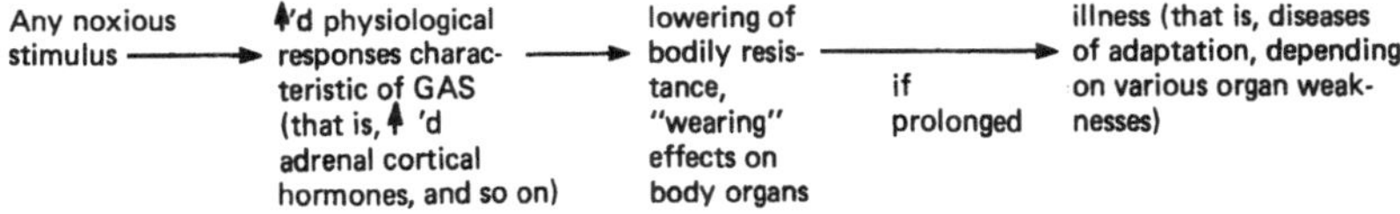

Figure 1. Selye's General Adaptation Syndrome model. From "Personality, Stress, and the Development of Physical Illness" by F. Cohen. In G. C. Stone, F. Cohen, and N. E. Adler (Eds.), *Health Psychology*, p. 83. Copyright 1980 by Jossey-Bass Publishers, San Francisco. Reprinted by permission.

Lazarus has defined four main modes of coping which he terms (1) information seeking, (2) direct action, (3) inhibition of action, and (4) intrapsychic or cognitive processes. He suggests that individuals differ in terms of their preferred coping styles, and also that they will utilize a variety of coping strategies, depending upon the circumstances in which they find themselves.

Lazarus also suggests, as have others (Hackett & Cassem, 1979), that some coping styles are more appropriate for some circumstances than for others. For example, Cohen and Lazarus (1973) found that patients who avoided or denied information about a forthcoming elective surgical operation showed faster and less complicated recovery from surgery than patients who sought information about their operation. The authors suggest that in a situation such as a hospital setting information seeking is not one of the most appropriate coping strategies because there is little a patient can do to better his own condition.

The idea that there are preferred styles of coping and that individuals will differ from one another on these preferences has relevance for any intervention strategy. There may be important interactions between the type of information presented to patients and the personality characteristics of that patient. If such factors are not taken into account, many intervention studies will show negative or inconsistent results.

Perhaps the best known study which attempted to pair nursing care procedures and personalities of patients was that of Cromwell, Butterfield, Brayfield, and Curry (1977). The authors assessed MI patients on

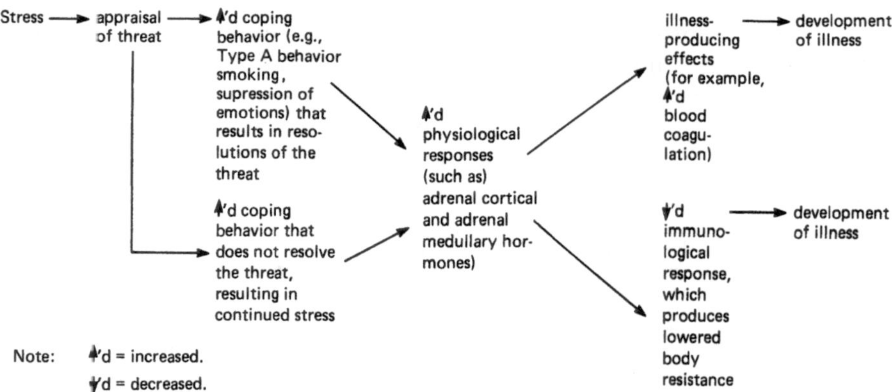

Figure 2. Cognitively mediated endrocrine/immunological mechanisms model. From "Personality, Stress, and the Development of Physical Illness" by F. Cohen. In G. C. Stone, F. Cohen, and N. E. Adler (Eds.), *Health Psychology*, p. 83. Copyright 1980 by Jossey-Bass Publishers, San Francisco. Reprinted by permission.

locus of control and found that except when internal locus of control was coupled with high anxiety, the external locus of control was unfavorable. They also investigated the extent to which information given to the patient was helpful and found that how this information was presented to the patient was important. If it was coupled with chores for him to do to foster his recovery, it was helpful. If it was not, then his recovery was hindered. Also, the patient's desire to receive the information could render it helpful or not.

RECENT LIFE EVENTS

Extending the knowledge that stress can have an adverse effect on health is the work on the stress associated with life events. It has been suggested (Engel, 1971; Holmes & Rahe, 1967) that there is a temporal relationship between the number of life events and the onset of illness regardless of whether the life event was a positive or negative one. The theoretical rationale behind this assumption is based on the evidence relating stress to illness. If an individual perceives change as stressful, or does not have adequate coping strategies with which to deal with change, then the change itself could contribute to the likelihood of illness developing. Exposure to this type of stress is not thought to cause the disease directly but will interact with genetic predisposition to disease or presence of a disease agent and alter the individual's susceptibility to these things (Selye, 1974).

A prospective study of Rahe and Arthur (1968) involved 2,500 American naval personnel between the ages of 17 to 30 who reported for duty on a 6-month cruise. They were asked to report the number of events that had occurred in their lives for the 6 months prior to the cruise. These events are presented on a 55-item Schedule of Recent Experience which provides scores for (1) number of events endorsed, (2) weighted scores termed Life Change Units (LCU), and (3) SLCUs or Subjective Life Change Units. For the naval personnel, when their responses were grouped into quartiles based on their precruise LCU scores, there was a significant relationship between number of LCUs and reported illnesses during the 6-month cruise.

This finding has been replicated by many other researchers, and Rabkin and Struening (1976) state:

> In both retrospective and prospective investigations, modest but statistically significant relationships have been found between mounting life change and the occurrence or onset of sudden cardiac death, myocardial infarctions,

> accidents, athletic injuries, tuberculosis, leukemia, multiple sclerosis, diabetes and the entire gamut of minor medical complications.

Garrity and Marx (1979) review the studies that have specifically related life events and coronary disease. They conclude that the evidence for the relationship is persuasive, although there is a disagreement in the literature regarding the lead time for these events and the experience of the illness. However, since many of the studies of victims of heart attacks have been retrospective (in the case of death based on the recollection of the surviving spouse), there have been methodological criticisms of their findings.

Dohrenwend and Dohrenwend (1978) discuss the research in the area of stressful life events and point out a number of difficulties inherent in making sense out of the research findings. They suggest that events must be looked at in the context in which they occur.

> The effect of a social change, or a change in interpersonal relations on the health of an individual cannot be defined solely by the nature of the change itself. The effect depends on the physical and psychological characteristics of the person who is exposed to the change and the circumstances under which it is encountered. (p. 93)

These authors are concerned about the items included in the Holmes and Rahe scale, with 29 of the original 43 listed events appearing to be symptoms or consequences of illness in and of themselves. They suggest that three distinct types of life events should be sampled in order to establish which ones have the most predictive utility. These three types would be those having to do with psychiatric events, those having to do with physical illness and injury events, and those which are independent of either the subject's physical or psychiatric condition.

Other criticisms of the Recent Life Events research has focused on statistical issues such as the small actual correlations between frequency of reported life events and occurrence of illness. These become significant because of the large samples utilized in the Holmes and Rahe research. There is also some concern over the fact that there appears to be an inverse relationship between age and number of life events reported, as well as a relationship between social class (Husaini & Neff, 1981), ethnicity, and reported life events. Lastly, the failure to include such relevant variables as social support networks and personality characteristics has been stressed (Rabkin & Struening, 1976; Dohrenwend & Dohrenwend, 1978).

Rahe himself points out that in his original sample of naval personnel the relationship between life events and illness held true only for approximately one-third of the group. He suggests that interest should focus on those individuals for whom life events appear to be detrimental in terms of health. He has done research in which he attempts to determine some of the underlying physiological mechanisms which differentiate these persons from those who are unaffected by life changes.

LIFE EVENTS AND CORONARY BYPASS PATIENTS

For our patients the Recent Life Changes Scale proved to be informative in a number of ways. First, there was a significant correlation between the number of life events occurring within the 6 months prior to surgery reported in patients' scaled scores on the life events scale and depression. In other words, patients who endorsed higher numbers of life events and had higher scaled scores were more depressed prior to surgery. This is consistent with our expectations and the findings of others (Billings & Moos, 1982; Johnson & Sarason, 1978) that number of life events and depression are related.

Second, the scores obtained prior to surgery on the life events scale were also predictive of depression 12 weeks after surgery, and this relationship was still true at 3 years. A third finding was that the scores taken at 12 weeks after surgery (that is, the number of life events occurring in the 12 weeks after surgery) were significantly related to the Beck Depression scores 3 years later. It appears that recent life events are predictive of depression not only prior to surgery but shortly after surgery and even 3 years later. Life events occurring shortly after surgery can also be predictive of later depression, and this can still manifest itself as long as 3 years later.

Crisis intervention appeared to affect this relationship between recent life events and depression. There is evidence that frequency of life events or changes prior to surgery can lead to depression in patients immediately after surgery. However, this effect appears to be neutralized if the patients receive crisis intervention.

The impact of crisis intervention was also evident on the interim measure of recent life events. For example, we looked at the subjective totals on recent life events in the period immediately after surgery and related this to depression 3 years later. At 3 years after surgery there was a tendency for patients who had a high subjective total of life changes 12 weeks after surgery to be less depressed if they were in the crisis intervention group. On the other hand, those who had a low subjective total of life events at 12 weeks and were in the control group remained about the same on their depression scores.

Finally, as we will see later, the number of recent life events in conjunction with other medical variables provides good predictive evidence for the efficiency of crisis intervention. This measure is a robust predictor in the case of middle-aged, male coronary bypass patients.

LOCUS OF CONTROL

The suggestion that locus of control has importance as an intervening variable in illness is based on the argument that stress is less damaging to an individual when the individual perceives himself as capable of controlling the stress in some way (Lazarus, 1974; Weiss *et al.,* 1982). There are significant individual differences, however, in the extent to which persons perceive themselves as capable of controlling the events in their lives.

For example, Rotter (1966) developed a Locus of Control Scale which categorizes individuals as either internal or external in their thinking about control. *Internal* persons perceive events as determined in some way by their own behavior, whereas *external* persons perceive events as due to some external force such as luck or fate. It is reasonable to expect that the crisis of hospitalization and surgery will call upon increased efforts to cope in patients and that these efforts are likely to be particularly strong for patients who have a need to control their environment. For them it might be assumed that they will find hospitals a particularly distressing setting because an individual can take so little responsibility for his own welfare.

The consequences of hospitalization are expected to vary from patient to patient. If the patient is external, his perception of lack of power will be reinforced in the hospital. If a patient is internal, his feelings of need for control and competence will be seen as interfering with his hospitalization. The external, passive patient may be seen as a good patient, whereas the patient who is internal may be seen as deviant and uncooperative; in short, in hospitals self-reliance and independence are frowned upon (Lorber, 1975; Taylor, 1979).

These intuitive speculations have been formally investigated, and the relationship between locus of control and distress in hospital has been verified. Various studies have indicated that locus of control is related to the adjustment of individuals to a hospital setting. For example, Seeman and Evans (1962) determined that internals react more favorably to the hospital setting and feel less alienated, whereas externals are more alienated and less satisfied, and suffer from extreme feelings of powerlessness and helplessness. They believe that the hospital structure promotes a general external view of control.

Lowery, Jacobsen, and Keane (1975) saw the patient who was about to undergo surgery as perceiving hospitalization as a totally uncontrollable situation. The road to health was under the control of "powerful medical others" and the patient's role was to remain passive throughout the hospital period. They found that patients who were higher on the external locus of control were more anxious in such a setting than were internals.

Auerbach (1979) reviews the literature on preoperative preparation for surgery and summarizes his remarks by suggesting,

> In summary, the manner of information presentation and the general dimension of perceived control are two independent variables that require further exploration. Of particular interest is the nature of the interaction between perceived control and the individual difference variables such as locus of control orientation or preferred coping style as they affect recovery. (p. 349)

Auerbach (1973) found that externals tend to have higher levels of anxiety prior to and after surgery, as well as a slower recovery rate. Also, externals knew less about their illness and/or surgery and asked fewer questions concerning their malady. They were also generally more passive. In a similar study, low-competence patients (externals with poor coping skills) were rehospitalized 62% of the time as a result of illness related to stress; high-competence patients (internals with adequate coping skills) were rehospitalized only 30% of the time.

Smith (1970) proposed that crisis patients would be more externally oriented than noncrisis patients because of the external stresses at the time of the crisis. After 6 weeks of crisis resolution it was found that those patients involved with crisis intervention shifted from an external orientation to an internal, and no such change occurred in the control group. In our study we observed similar results, with the scores on the Locus of Control Scale going down from an average of 7.37 to an average score of 6.22 in our patients who received crisis intervention. For those patients who did not receive counseling, the locus of control scores remained the same (7.37–7.80) 12 weeks after surgery.

A study which grouped patients according to their scores on the Locus of Control Scale and then manipulated their hospitalization experience was conducted by Cromwell *et al.* (1977). In this study myocardial infarction patients were given the latitude to activate their own cardiac monitors to secure an EKG tracing when they experienced discomfort (and to engage in various exercises under close staff supervision), and other patients were given only standard bed rest. For internal locus of control subjects, high levels of participation in conjunction with specific information resulted in short hospital stays, whereas high-participating subjects who had received general information only had relatively long hospital stays.

Finally, perhaps most relevant to our research are the findings that relate locus of control to depression and hopelessness. Since depression was our dependent measure both before and after surgery, we were interested in seeing to what extent the locus of control score of an individual related to the level of that individual's depression. O'Leary, Donovan, Cysewski, and Chaney (1977) compared groups of alcoholics on the Beck Depression Scale and Rotter's Locus of Control and found that subjects with an external locus of control and minimal control were significantly more depressed. Similarly, Prociuk, Breen, and Lussier (1976) related the results on the Beck Depression Scale and Rotter's Locus of Control Scale. They found that depression was significantly related to perceived external control.

As other research has found, and as theory would predict, those of our patients who were more internal in their locus of control were significantly less depressed than patients who were more external. On the Beck Depression Scale prior to surgery, those who were more internal had an average score of 7.05; at 12 weeks the average score was 6.94, and at three years it was 6.55. On the other hand, those who were more external had the following Beck Depression scores: prior to surgery 12.03, at twelve weeks 11.07, and at three years 10.77. An interesting exception to this relationship occurs with the subjects who are internal in their locus of control who report a high subjective total on the life events scale presurgically. For them, this subjective score is highly correlated with depression three years later.

Johnson and Sarason (1978) investigated the relationship between life changes, depression, and anxiety as a function of the subject's locus of control orientation. They assumed that life changes would have their most adverse effect on individuals who perceive themselves as having little control over environmental events and that there would be significant correlations between life changes and depression in those who were external in their locus of control orientation. They suggested that it was not only the number of life changes experienced by an individual that was aversive but also the extent to which an individual felt he had some control over life events. Their results support their hypothesis, and our results are in close agreement.

We found that patients who were classified as external locus of control presurgically and who endorsed a high number of recent life events also scored high on the Beck Depression Scale. This was true on the Beck at all three points in time (before surgery, 12 weeks, and 3 years) and the significance of this relationship became stronger over time.

A final note on the concept of locus of control and its relationship to illness behavior: Although there is literature on the use of Rotter's Locus of Control Scale with populations of ill and hospitalized patients, contemporary studies now tend to use one of the newer Health Locus of Control Scales (Lau, 1982; Lau & Ware, 1981). These scales are based on Rotter's original scale but include more categories. For example, Lau suggests that it is important to look not only at general locus of control but also at such things as health care attitudes, health status perceptions, and the value placed on health. He found that, contrary to earlier notions regarding the association between internal and external factors in determining health outcomes, beliefs in personal and provider control over health were positively associated. In other words, persons who believe that health outcomes can be controlled or influenced also tend to believe both in themselves and in the ability of doctors to affect those outcomes. Research using these scales has provided interesting information regarding health behavior of a variety of populations of subjects (Deaton & Olbrisch, 1981; DeVito, Bogdanowicz, & Reznikoff, 1982). However, our main interest was in looking at the relationship of personality to depression.

We were surprised to find that there appeared to be no interactions between locus of control and crisis intervention. Although external locus of control was related to depression, crisis intervention did not appear to affect differentially those patients who were external in their locus of control compared to those who were internal. We did note, however, that locus of control scores became lower in the treatment or crisis intervention group over time (7.37–6.22) and those in the control or no treatment group remained the same (7.37–7.80). Because lower scores signify more internal locus of control, and because internal locus of control signifies less depression, our treatment groups changed in the appropriate direction.

One of our reasons for not achieving statistical significance for this relationship may have been the homogeneity of our patient sample. Our average scores on the Locus of Control Scale were significantly lower than the average scores obtained by Rotter on his original sample (Rotter, 1983). This would signify that our patients were primarily internal in their locus of control and did not provide us with a wide range of scores. Although it is not surprising that middle-aged men in our society have become more internal in their thinking compared to the college students who made up Rotter's original sample, this lack of variance makes it difficult to achieve statistical significance.

TYPE A

It would not be feasible for a book about coronary bypass surgery to ignore the proliferation of literature linking Type A or coronary-prone behavior to increased cardiac risk. Ever since the term was coined by Friedman and Rosenman (1974), Type A has provided provocative evidence of the relationship between psychological variables and risk of heart disease. In our study we utilized the Jenkins Activity Scale prior to surgery and again on our 3 year follow-up study. We wished to assess the number of patients who would be classified as Type A or B in our sample, the extent to which Type A and B characteristics were related to depression, and the way in which this might interact with the beneficial effect of crisis intervention.

Much to our surprise, our patients did not show a prevalence of Type A responses on the Jenkins and were divided almost equally between Type A and Type B. This result is similar to the findings of Croog and Levine (1982), who followed heart attack victims for 8 years, and raises some questions regarding the utility of this concept in populations of older subjects.

Excellent reviews of the current status of the Type A concept have been written by Steptoe (1981), Chesney and Rosenman (1982), and Krantz, Glass, Schaeffer, and Davia (1982). This section will summarize their comments and also include research studies that appear to be relevant to our experience. The research evidence suggests that the importance of the Type A concept in the field of coronary heart disease cannot be overemphasized.

Recently the National Heart, Lung, and Blood Institute assembled a review panel of 50 eminent scientists who represented a wide range of biomedical and behavioral specialities. Their task was to critically examine the evidence from all of the studies relating the Type A behavior pattern to the incidence of coronary heart disease. Their final report opened with the following statement:

> The review panel accepts the available body of scientific evidence as demonstrating that Type A behavior, as defined by the structured interview used in the Western Collaborative Study, the Jenkins Activity Survey, and the Framingham Type A behavior scale, is associated with increased risk of clinically apparent CHD in employed, middle-aged, U.S. citizens. This risk is greater than that imposed by age, elevated values of systolic blood pressure and serum cholesterol, and smoking, and appears to be of the same order of magnitude as the relative risk associated with the latter three of these factors. (Chesney & Rosenman, 1982)

MEASUREMENT OF TYPE A BEHAVIOR

The original way of assessing Type A was on the basis of a structured interview (Friedman & Rosenman, 1974), which was tape-recorded and later assessed by a trained rater. Subjects were assessed on the basis of their responses to questions which were intended to elicit evidence of competitive drive, impatience, and irritations. Later the Jenkins Activity Survey, a self-administered questionnaire which yields a continuous distribution of scores, was developed.

Originally, Type A was conceptualized as an individual personality trait which interacts with environmental stress; however, now it is thought of as a characteristic way of responding or coping with a variety of environmental stressors. Initially, Type B behavior was considered simply the absence of Type A behavior, whereas now there appears to be evidence that it is a behavior pattern of its own. Type B's have been found to be more introverted, relaxed, deferential, and unhurried. This is in contrast to Type A's, who are characterized by excessive competitive drive, impatience, hostility, and accelerated speed and motor movements.

RELATIONSHIP OF TYPE A TO CORONARY HEART DISEASE

Evidence for the relationship of Type A behavior to coronary heart disease (CHD) originally came from the Western Collaborative Study (WCGS), which began in 1960–61. This California project traced over 3,000 employed middle-aged men for 8½ years. Although half of the population was classified as Type A and half as Type B on initial interview, those who had heart disease were overrepresented in the Type A category. During the initial screening, 113 cases of coronary heart disease were discovered; 68% of those in the 39–49 age group were Type A, whereas 73% of those in the 50–59 age group were in this category. During the years of the study, 257 persons developed coronary disease, of whom 50 did not survive. In both younger and older age groups, the rate of heart disease in Type A's was twice that of Type B's. This was true even when other factors such as serum cholesterol, blood pressure, and cigarette smoking were controlled for.

Participants in the WCGS were evaluated on both the Structured Interview and the Jenkins Activity Survey for incidence of Type A characteristics. The Jenkins was also successful in predicting incidence of heart disease but not so successful as the structured interview. Because the sample was so large, the smaller numbers became statistically significant. It should be noted that studies have repeatedly found the structured interview developed by Friedman and Rosenman to be superior to the Jenkins Activity Survey in classifying Type A and Type B behavior critical for later heart disease. However, because of its ease of administration, the Jenkins continues to be the instrument of choice for many researchers.

Further support for the ability of a questionnaire to identify Type A characteristics came from the Framingham study, which was conducted between 1965 and 1967 and which included both men and women. In this study 1,822 male and female subjects were administered an extensive questionnaire from which the Framingham Type A scale was developed. Of this larger sample, 1,674 subjects between ages 45 and 70 years who were free of CHD were followed for 8 years. The FTAS showed a significant correlation with the prevalence of CHD in both sexes, and this was true even when other risk factors were controlled.

Recently studies in other countries utilizing adaptations of the Type A scale have indicated a relationship between Type A subjects and later CHD. The Bortner scale has been used in the large Belgian Disease Prevention Project (Kornitzer, Magotteau, Degre, Kittel, Struyven, & VanThiel, 1982) and recently (Appels, Jenkins, & Rosenmann, 1982) in a similar study in Holland administered the Jenkins Activity Survey to 2,712 Dutch men. In this research the Dutch adaptation of the scale identified correctly 73% of those who later suffered angina pectoris.

In spite of these findings, it has been speculated that Type A behavior is only evidenced in competitive, highly industrialized societies with values similar to those of the United States. Research with Japanese-Americans has suggested that *hard-working* can be differentiated from *hard-driving/competitive* and that the former is not related to increased incidence of CHD although the latter tends to be. A recent study done by Stensrud, Gilgen, and Koloc (1982) assessed subjects on their belief systems. They found that there was a negative correlation between incidence of Type A responses in those whose belief systems were more Eastern compared to subjects who were more Western in their beliefs.

Steptoe (1981) comments on the differences in the incidence of Type A behavior in different populations and notes that not all environments and cultures elicit the response pattern. He points out also that there have been consistent positive correlations between Type A and socioeconomic status as well as educational achievement. Type A scores have been shown to vary directly with job level, and blacks tend to score lower than whites. It has also been observed that ratings may decline in late middle age when occupational goals have been achieved. The coronary-prone behavior pattern is found more often among employed than unemployed women but only in higher education groups. In view of the observed relationship between Type A characteristics and economic and educational success, it is hard to describe it as a socially undesirable response pattern. Steptoe also points out that in the WCGS study 74% of the men at highest risk on all factors related to CHD continued to remain free of the disease for over the 8½ years of the study.

TYPE A AND THE SYMPATHETIC NERVOUS SYSTEM

Currently the greatest interest in Type A behavior and coronary risk is focused on the possibility that this relationship is mediated in some way by a heightened responsiveness of the sympathetic nervous system when confronted with a challenging task (Glass, 1977). In research with Type A's and B's who were presented with impossible tasks, it was demonstrated that initially Type A's responded with hyperresponsivity after which they became depressed and were noted to underrespond. This was in contrast to Type B's, who continued to respond at the same level. Glass has proposed that Type A's are particularly vulnerable to the effects of these alternating acute phases of active coping and giving up: "The greater likelihood of the disease in Type A might be explained in terms of the cumulative effects of the excessive rise and fall of catecholamines released by the repetitive interplay of pattern A and uncontrollable stress."

Ditto (1981) measured autonomic responding in Type A's and B's in two types of tasks. One type had a time limit and the other had no deadline. He found that during the no-deadline condition Type A's maintained their activity levels and slightly increased the number of problems they attempted compared to the deadline conditions, whereas Type B's slightly decreased their activity levels and numbers of problem they attempted compared to the deadline condition. In spite of the behavioral differences between the two conditions for Type A's, however, there were no corresponding changes between the two conditions on any of the autonomic measures.

Snow and Glass (1981a) looked at the differential autonomic reactivity of Type A's and B's on tasks which were congruent with and incongruent with their favored response styles. Although there were no differences in performance or number of problems solved between the two groups, levels of blood pressure and heart rate varied depending upon the condition. Systolic blood pressure was significantly higher in A's during the slow condition and was higher for B's during the fast condition. A's and B's were not significantly different during the very fast condition. Heart rate for A's was higher in the slow condition and for B's in the fast condition, and again no differences were found in the very fast condition. The authors suggest that the Type A variable is an interaction between personality style and situational factors, rather than a personality trait which is exhibited in all settings.

In a later study (1981b) the same authors looked at coping strategies used by Type A's and B's when confronted with tasks which were incongruent with their typical response style. Type A's used more defenses when in tasks which were challenging such as very fast and slow conditions. The types of defenses used by Type A's were more likely to be primitive ones such as denial or suppression, whereas Type B's utilized more advanced defenses such as intellectualization. The authors suggest that the use of primitive defense patterns provides evidence beyond the cardiovascular evidence that Type A's find incongruent settings aversive.

Graham, Ho, Thoresen, *et al.* (1981) looked at the extent to which Type A's suppressed fatigue on a treadmill task. They found that the more Type A an individual was, the more likely he would be to deny his fatigue when tested on a treadmill task. On the basis of information that Type A's exhibit heightened physiological arousal under stress and also that they suppress fatigue, Strube and Werner (1981) conjectured that they might be less attentive than others to their own internal physiological states. They attempted to teach Type A's and B's to control their heart rates and found that they differed in this ability. Although these efforts were made during deadline and no-deadline conditions, it was found that Type A's were only successful under the deadline condition.

The effect of competition on physiological arousal in Type A's has been investigated by a number of writers. Gamino and Houston (1981) looked at the effect of severe to mild failure with and without feedback on a variety of physiological responses. They found that participation in these competitive conditions resulted in increased physiological arousal compared to a control or no-stress condition. They felt that these findings provide support for the hypothesis that Type A's react with greater autonomic arousal when confronted with competitive challenge.

Dembroski (1983) has conducted a number of studies relating heightened physiological responding to Type A's in challenging tasks. He finds that catecholamine activity increases more in Type A's when they are placed in impossible tasks. Not only do Type A's show greater reactivity but they stay on the task longer. When working on impossible tasks under harassment the blood pressure of Type A's shows greater increases than that of Type B's, although Type B's show blood pressure increases as well. In his studies, Dembroski finds that it is possible to differentiate Type A's who will react most to challenge by their voice styles in an interview. Those with the most strident voices show the most reactivity during high challenge.

Dembroski compared men and women on challenge tasks by pairing males with males, males with females, and females with females. His research shows that both males and females show typical Type A responses but males are always most reactive whether paired with males or females. When their blood pressure was compared during final exams, both males and females reported equal subjective stress but males showed higher blood pressure reactivity. Males had higher blood pressure to begin with and it got higher during competition. Male heart rate was lower initially but got higher during competition.

Perhaps the most striking finding regarding the heightened tendency for physiological responding comes from Kahn, Kornfeld, Frank, Heller, and Hoar (1980). They looked at blood pressure of patients undergoing coronary bypass surgery and found that there were significant correlations between the interview ratings of patients as Type A and systolic blood pressure rise during surgery. These data suggest that patients with Type A behavior characteristics manifest an autonomic hyperactivity which is present under general anesthesia.

Williams, Lane, Kuhn *et al.* (1982) report that they have begun to differentiate some of the specific physiological responses of Type A's in relation to specific situational stressors. They found that Type A subjects showed greater muscle vasodilation and more secretion of norepinephrine, epinephrine, and cortisol than Type B subjects during mental work. During sensory intake, however, they found the Type A's hyperresponsive for testosterone and, among those subjects with a positive family history of hypertension, for cortisol. They believe these findings suggest that young Type A men are not generally hyperresponsive to environmental challenge but are so only in certain physiological responses and only as a function of the type of challenge and genetic predisposition.

In spite of the strength of the relationship between Type A behavior and the incidence of coronary heart disease, the concept is not without its critics (Radley, 1982). It is often pointed out that although Type A is not intended to be considered a personality style but rather a characteristic way of responding under certain conditions, it is often used in the research as though it were. This creates confusion over how to modify or correct for Type A, because it is difficult to know whether to emphasize changes in the individual or changes in the environment, or both. For example, Radley makes the following points: If Type A is to be considered a suitable subject for medical intervention, then the emphasis will have to be on its physiological variables.

> If coronary heart disease develops as a result of certain cultural organizations or individual styles of living, then little in the way of technical intervention by medical practitioners will be likely to affect it greatly. There was once a brief literature on the psychological features which might predispose people to pulmonary tuberculosis. The discovery of the tuberculosis bacillus and the effective use of vaccination meant that this approach was promptly rendered sterile for the purpose of curing TB. (p. 113)

Several authors (Deszca & Burke, 1981; Roskies, 1982) have discussed the problems associated with modifying Type A behavior. Roskies points out that one of the biggest problems is the fact that Type A behavior is economically advantageous and has a very high incidence in our competitive society. Since there are so many successful men who are Type A and who do not develop coronary heart disease, it does not make economic sense to consider attempting to change this. For example, prevalence rates for Type A have been estimated from 50% to 75% in North America among symptom-free individuals.

One could argue that because Type A appears only to result in increased autonomic nervous system responding when the individual is under stress, it would be wise to remove Type A's from stressful situations. The problem is in deciding which circumstances are stressful and which are not. The research of Snow and Glass is a good example of the extent to which Type A's showed physiological arousal under slow conditions while their Type B counterparts showed the same physiological arousal under fast conditions. In other words, what is stress for a Type B may not be stress for a Type A.

However, it is generally agreed (Dembroski, 1983) that it is not the increased physiological hyperreactivity of the Type A's which poses the medical danger but their propensity to manifest longer-lasting rises in autonomic and neuroendocrine levels once the stress has been removed. A goal of intervention, therefore, might be that of reducing the frequency, intensity, and duration of episodes of arousal.

Roskies (1979) has completed a study with 66 healthy, middle-aged, middle-level managers in Canada who volunteered for a 13-session stress management program sponsored by the large corporation in which they work. These men participated in two types of intervention strategies intended to modify their Type A behavior; they were also assessed for psychological state, blood pressure, urinary catecholamines, and serum cholesterol as well as testosterone. The two approaches (psychotherapeutic or behavior therapy) produced desired results in terms of changes in serum cholesterol, systolic blood pressure, diastolic blood pressure, and serum triglycerides; and these alterations were still in evidence at the follow-up 6 months after completion of treatment. However, when Roskies's data are examined clinically (Suinnn, 1982) the results can be interpreted differently. Although both psychotherapy and behavior therapy groups achieved statistically significant reductions in cholesterol, only the behavior therapy group dropped to a low clinical risk level. Those in the psychotherapy group, although showing a drop in cholesterol, were still at an intermediate risk level.

Friedman, Thoresen, Gill *et al.* (1982) studied 1,035 consecutive post-infarction patients to determine the feasibility of altering Type A behavior and to see what effect this alteration would have on subsequent rates of infarction and cardiovascular death. Three hundred subjects enrolled in small groups and received cardiologic counseling on the usually accepted coronary risk factors. Six hundred subjects received, in addition to this, advice and instructions on how to diminish the intensity of their Type A behavior. The remaining subjects were controls and received no counseling other than their annual medical examination and interviews. Ninety-eight percent of the subjects had been assessed as moderate to severe Type A on the basis of their videotaped structured interview at the beginning of the study. After the first year of this 5-year study, the rates of infarction and cardiovascular death were lower among subjects who received both cardiologic and behavioral counseling than among control subjects. Also, the rate of nonfatal infarction was lower among subjects who received behavioral counseling than among those who received only cardiologic counseling or those who dropped out of either counseling group.

These recent research findings appear to be promising for they are concentrating on populations at clinical risk for heart attack and are combining physiological measures with psychological assessments. The next section will compare our patients to research studies which have interested themselves in similar groups of subjects.

A study by Byrne (1981) has suggested that Type A individuals may organize their styles of living in such a way as to increase the probability of encountering stressful life events. The suggestion is put forward that Type A is, in essence, a pattern of response tendencies which dictate the organization of life style. Persons who are ambitious, impatient, and goal-directed in the employment situation are likely to organize their lives in such a way as to facilitate the expression of these characteristics. Moreover, they are more likely than others to place themselves in social, occupational, or environmental positions which maximize the likelihood of encountering life events already associated with myocardial infarction research. Byrne found a significant correlation between frequency of life events in the previous 12 months and a myocardial infarction. Interestingly, our study with heart patients found frequency of recent life events related to poor psychological outcome after heart surgery and poor psychological attitude prior to coronary bypass surgery.

Because our patients were evenly divided between Type A and Type B, our findings do not support those which suggest a relationship between Type A and coronary heart disease. Others, however (Croog & Levine, 1982), have also noted a similar distribution in their populations of heart attack victims. One possible reason for this discrepancy may be the age of our patients. Most of the research on Type A or coronary-prone behavior has been conducted on younger populations of middle-aged working American males. Our patients were older and many of them were retired prior to having coronary bypass surgery. This would be expected to reduce the incidence of Type A among this group. Sparacino (1979) looked at Type A aging and mortality and suggests that the clinical significance of Type A behavior may shift radically in old age. He points out that Type A score is inversely related to age for both men and women. Although the relationship between Type A and CHD is true of younger groups, for persons over 65 the relationship is insignificant.

In spite of the fact that we did not have a preponderance of Type A's in our population, we did find a significant effect for the Job Involvement subscale of the Jenkins and the extent to which crisis intervention helped depression. Twelve weeks after surgery, for those patients who scored high on the Job Involvement component of the Type A scale participation in crisis intervention appeared to be more help. This result is interesting in view of the research by Vickers, Hervig, Rahe, and Rosenman (1981), who discuss the relevance of Job Involvement to coronary heart disease. They suggest that the Job Involvement subscale measures the individual's coping skills compared to the other subscales of Hard Driving and Speed and Impatience. They found that Job Involvement showed a high relationship to coping and low relationship to defensiveness whereas Speed and Impatience was related to high defensiveness and Hard Driving to low coping scores. In terms of crisis intervention, which helps individuals recover their coping skills, high scores on Job Involvement would be predicted to be favorable predictors of good utilization of counseling. This appears to be the case for our patients, because those Type A's high on Job Involvement were those who benefited most from the crisis intervention at 12 weeks (see Appendix A in Chapter 7).

PSYCHOLOGICAL VARIABLES AND THEIR RELATIONSHIP TO MEDICAL VARIABLES

SCL-90 R

Chapter 7 provides a detailed account of the statistical analyses applied to our data. This chapter, like Chapter 8, has attempted to fit the statistical findings into some kind of conceptual and theoretical framework based on the existing literature on coronary heart disease and coronary bypass surgery. A finding that has not been discussed is that of the decrease in General Symptom scores on the SCL-90-R which were noted in both treatment and control patients after surgery. These scores were also important contributors to the formulae for prediction of depression in conjunction with medical variables. Our average scores did not approach clinical significance in terms of the usual cutoff score for interpretation of the SCL-90-R. However, they obviously provide a global measure of psychological distress, which relates to the impact of coronary bypass surgery.

For surgeons who are interested in obtaining a global score of symptom severity, perhaps the SCL-90-R would be a useful tool. The fact that the Analogue (completed by the patient's spouse) provided such a good estimate of depression in the patient would also suggest that the SCL-90-R could be included in a surgical test battery. Since it is expected that patients will improve in their average score on the SCL-90-R, scores which are in the high range after surgery indicate a possible emotional problem that has not been resolved.

PATIENT SELECTION FOR CRISIS INTERVENTION

We reported in Chapter 7 that three presurgical factors stood out in their ability to predict statistically the effectiveness of crisis intervention on depression in our patients. These were (1) the presence or absence of left main coronary disease, (2) the number of bypasses, and (3) the number of events endorsed on the Recent Life Changes Scale expressed as a predictive equation. Using this equation we have selected two case histories which fit these criteria, one in the treatment and one in the control group. Again, the reader is reminded that this equation was derived after the completion of the study and that patients were not selected for crisis intervention this way. It remains necessary to conduct a prospective study in which this is done.

The instructions for selecting patients which would benefit from coronary bypass surgery require the surgeon to look first at the presence or absence of left main coronary artery disease. If it is absent, the patient is likely to benefit from crisis intervention. If it is present, the next criterion to consider is the number of bypasses. If there are three or more bypasses the patient should be considered for crisis intervention. Finally, in the presence of left main coronary artery disease, with fewer than three bypasses the number of recent life changes becomes the deciding factor.

CASE HISTORIES

We selected two cases, one from our treatment group and one from our control group, on whom we had 3-year follow-up information. We will present the treatment group case first.

Patient X (Treatment Group)

	Before test	12 weeks	3 years
SCL-90-R GSI	80	81	81+
Beck	17	12	13
Locus of Control	8	14	14
Analogue of Depression	—	70	51
Recent Life Changes	12	3	—
Type A			

Patient X qualified for crisis intervention because he did not have left main coronary artery disease. However, he also demonstrated other predictive characteristics as he had four bypasses and endorsed a high number of recent life changes. Medically, he had a history of prior myocardial infarct, hypertension, family heart disease, and angina. He did not have triple coronary disease; he had not had previous heart surgery; his blood pressure was 120/70; his ejection fraction was 55%; and he was New York Heart Association Class 4. He is almost identical in medical characteristics to Patient Y with whom he will be compared.

Psychosocial factors in the background of Patient X include the fact that he was 51 years old, married, with some family living in the community. He had lived in Miami for over 30 years and had been employed in a skilled trade until his heart condition forced him to retire. His scores on the psychological measures prior to surgery classified him as depressed with a Beck score of 17, internal in his locus of control, and significantly upset emotionally, evidenced by his score on the SCL-90-R. He endorsed a high number of recent life changes and reported a high subjective rating of these events as well.

This patient had a difficult time with the crisis of coronary bypass surgery. His wife was a nurse, and this fact appeared to increase his distrust of the medical personnel and hospital staff. One of the functions of the crisis intervention counselor is that of maintaining communication with the hospital team and the family. However, in this case, the patient's wife assumed this role. Whether her professional training served to threaten the medical staff, or whether her expectations were unrealistic, her involvement appeared to create more distrust and psychological concern for the patient.

The crisis intervention counselor found the patient and his wife a difficult couple to work with. They appeared to be experiencing significant marital stress and during the counseling sessions often raised the issue of divorce as a real possibility. The patient continued to exhibit depression throughout the period of crisis intervention and asked the counselor if she could give them extra appointments. Unfortunately, because this was a research study and not a clinical project, this request had to be denied. In order to make a valid comparison between patients, we were committed to keeping the number of counseling sessions identical for all participants. It is difficult to know what the outcome might have been if crisis intervention could have been intensified

The patient and his wife had limited financial resources, and he continued to experience medical symptoms that exacerbated his depression. Because of the nature of the patient's work, his illness posed a real threat to his future job security. In addition, one of their children was exhibiting behavior problems in the community, and the parents differed sharply over the correct way to handle the discipline situation.

At the end of the crisis intervention period the counselor writes,

> I requested permission to get in touch with this patient's physician in order to suggest that they receive further supportive counseling after I terminated with them. In spite of the fact that I did not feel I was accomplishing very much because of the magnitude of their problems, there were tears in their eyes at our last session. They had obviously found the sessions more meaningful than I had realized, and I honestly don't feel comfortable leaving them at this time. They have just begun to recognize where some of their difficulties are coming from and I feel frustrated that this has to be my last visit with them.

Probably crisis intervention provided under normal clinical circumstances would not have terminated at this time but would have continued until the family had been restored to an adequate level of coping. Three years later, this patient and his wife were interviewed by an independent counselor who provided us with the following assessment of their situation.

The patient appeared still to feel that his medical outcome was not as satisfactory as he had hoped. In spite of this, however, he was working part time, whereas he had been unable to work prior to the surgery. He also participated in social activities, something he was unable to do before his operation. He felt that his life had changed significantly since the bypass operation in a number of respects and felt that his biggest worry still was the unpredictability of his health. In spite of his statements that he had a good relationship with his doctor and that his doctor provided him with support and information, he still trusted his wife's opinion more when it came to medical matters.

When asked if they had found the crisis intervention helpful, both husband and wife agreed that it had meant a lot to them to have someone with whom to talk. The interviewer noted that the patient appeared to derive a great deal of support from his religious beliefs, that he still demonstrated some distrust of doctors, but that he and his wife seemed to feel comfortable about openly expressing their feelings. His main worry seemed to be that of concern over his ability to continue to work steadily and the unpredictability of his health with regard to this issue.

Although this case does not appear to be a total success in view of the extent of depression still evident, it would be hard to imagine the condition of this patient if crisis intervention had not been provided. He appears to have less depression now and his point of view on the Locus of Control Scale appears to have moved more in the external direction. This has probably helped him use his religious beliefs for comfort. When the notes of the crisis intervention counselor 12 weeks after surgery are compared to the interviewer's impression 3 years later, it does appear that this patient has recovered a reasonable emotional balance and is functioning better than might have been expected.

Patient Y (Control Group)

	Before test	12 weeks	3 years
SCL-90-R GSI	59	52	53
Beck	7	14	8
Analogue of Depression	—	0	5
Locus of Control	8	6	12
Recent Life Changes	7	6	—
Type B			

According to our formula, this patient would have qualified for crisis intervention in spite of the fact that he had left main coronary artery disease because he had three bypasses. He is similar to Patient X in almost all medical aspects, having had a family history of heart disease, a prior myocardial infarction, and a history of hypertension. He did not have triple coronary disease; he had not had previous heart surgery; he had a history of angina, blood pressure readings of 120/70, an ejection fraction of 53%, and was a New York Heart Association Class 4.

Psychologically, however, this patient differed from Patient X on a number of dimensions. He was classified as Type B on the Jenkins Activity Scale; he did not show signs of significant psychological distress on the SCL-90-R; and he did not indicate a large number of recent life changes. His responses on the Locus of Control Scale, however, showed him as being initially internal and them becoming more external, and his score on the Beck Depression Inventory was significantly depressed 12 weeks after surgery.

This patient was 65 years old and had moved to Miami during the 5 years prior to his heart surgery after a long career as a salesman. He was married and most of his family lived in this community. He attributed his heart problems to stress associated with work and with his personal life and felt that the surgery had a negative effect on his family relationships.

The patient was not part of the crisis intervention group; therefore, there is less information available about his progress other than the interview at 12 weeks and the report of the independent interviewer 3 years after surgery. Twelve weeks after surgery, the patient stated he would not undergo surgery of this type again. On the contrary, he was persuaded that he could have managed his angina successfully with a careful program of diet and exercise. He expressed very negative feelings toward the doctors and the hospital and the amount of money his surgery had cost him.

Three years later, the patient had changed his mind slightly. He still stated that he would advise others to pursue all possible alternatives to surgery but to accept it as a last resort. He himself continued to take the same amount of medication as he had prior to surgery and still attempted to follow the Pritiken diet. He perceived the dietary restrictions as aversive, however, whereas shortly after surgery he believed that the diet would have been a preferable alternative. The biggest single complaint of this patient was his generally depressed frame of mind since surgery. He described himself as having a changed personality and felt that he was a "dull vegetable" compared to his former self. He considered his biggest difficulty during recovery to be self-pity, and he remembered a period of approximately 1 month following surgery when he felt very down and sorry for himself.

The independent interviewer who spoke to him and his wife after an interval of 3 years described him in the following way:

> This patient states that his personality has changed since surgery. Everything in his life appears more bland. He was able to describe his feelings of self-pity during the period immediately following surgery and still seems somewhat dissatisfied with his present lifestyle. My impression is that crisis intervention for this patient might have helped.

Medical variables and psychological variables appear to intertwine in complex ways to result in either positive or negative physical and psychological outcomes after coronary bypass surgery. It is an interesting possibility that utilization of a simple formula, coupled with a standardized psychological assessment, can identify suitable candidates for the crisis intervention approach. Because depression appears to have such a powerful influence on the patient's capacity to return to a productive life, and because it can affect his perception of his physical well-being, it is to be avoided if at all possible. We have found crisis intervention for appropriate candidates to be capable of either eliminating the presence of depression entirely or ameliorating its effects when it occurs.

REFERENCES

Appels, A., Jenkins, C. D., & Rosenman, R. H. Coronary-prone behavior in the Netherlands: A cross-cultural validation study. *Journal of Behavioral Medicine,* 1982, *5* (1), 83–90.

Auerbach, S. M. Trait–state anxiety and adjustment to surgery. *Journal of Consulting and Clinical Psychology,* 1973, *40,* 264–271.

Auerbach, S. M. Preoperative preparation for surgery: A review of recent research and future prospects. In D. J. Oborne, M. M. Gruneberg, & J. R. Elser (Eds.), *Research in psychology and medicine* (Vol. 2). New York: Academic Press, 1979.

Billings, A. C., & Moos, R. H. Stressful life events and symptoms: A longitudinal model. *Health Psychology,* 1982, *1* (2), 99–117.

Blachly, P. J., & Blachly, B. J. Vocational and emotional status of 263 patients after heart surgery. *Circulation,* 1968, *38,* 524–532.

Brown, J. S., & Rawlinson, M. Relinquishing the sick role following open-heart surgery. *Journal of Health and Social Behavior,* 1975, *16,* 12–27.

Brown, J. S., & Rawlinson, M. The morale of patients following open-heart surgery. *Journal of Health and Social Behavior,* 1976, *17,* 134–144.

Byrne, D. G. Type A behaviour, life-events and myocardial infarction: Independent or related risk factors? *British Journal of Medical Psychology,* 1981, *54,* 371–377.

Chesney, M. A., & Rosenman, R. H. Type A behavior: Observations on the past decade. *Heart and Lung,* 1982, *11* (1), 12–19.

Cohen, F. Personality, stress, and the development of physical illness. In G. C. Stone, F. Cohen, & N. E. Adler (Eds.), *Health psychology.* San Francisco: Jossey-Bass, 1980.

Cohen, F., & Lazarus, R. S. Active coping processes, coping dispositions, and recovery from surgery. *Psychosomatic Medicine,* 1973, *35,* 375–389.

Cromwell, R. L., Butterfield, E. C., Brayfield, F. M., & Curry, J. J. *Acute myocardial infarction: Reaction recovery.* St. Louis: Mosby, 1977.

Croog, S. J., & Levine, S. *Life after a heart attack: Social and psychological factors eight years later.* New York: Human Sciences Press, 1982.

Deaton, A. V., & Olbrisch, M. R. *Health perceptions as a function of illness and hospitalization experiences.* Paper presentation at the 89th annual meeting of the American Psychological Association, Los Angeles, August 24, 1981.

Dembroski, T. *Stress, cardiovascular reactivity and coronary prone behavior.* Paper presentation at the Miami Symposium on Stress and Coping, a Research Symposium, Miami, Florida, February 21–22, 1983.

Deszca, G., & Burke, R. J. Changing Type A behaviour. *Canadian Psychology,* 1981, *22* (2), 173–187.

DeVito, A. J., Bogdanowicz, J., & Reznikoff, M. Actual and intended health-related information seeking and health locus of control. *Journal of Personality Assessment,* 1982, *46* (1), 63–69.

Ditto, B. *The Type A behavior pattern, time-limited performance, and autonomic response.* Paper presented at the annual meeting of the American Psychological Association, Los Angeles, August 1981.

Dohrenwend, B. S., & Dohrenwend, B. P. Some issues in research on stressful life events. *Journal of Nervous and Mental Disease,* 1978, *166,* 7–15.

Engel, G. L. Sudden and rapid death during psychological stress: Folklore or folk wisdom? *Annals of Internal Medicine,* 1971, *74,* 771–782.

Friedman, M., & Rosenman, R. H. *Type A behavior and your heart.* New York: Knopf, 1974.

Friedman, M., Thoresen, C. E., Gill, J. J., Ulmer, D., Thompson, L., Powell, L., Price, V., Elek, S. R., Rabin, D. D., Breall, W. S., Piaget, G., Dixon, T., Bourg, E., Levy, R. A., & Tasto, D. L. Feasibility of altering Type A behavior pattern after myocardial infarction. *Circulation,* 1982, *66* (1), 83–92.

Gamino, L. A., & Houston, B. K. *Competitive failure and the Type A coronary-prone behavior pattern.* Paper presentation based on first author's doctoral dissertation, Texas A&M College of Medicine, Scott and White Clinic, Temple, Texas, 1981.

Garrity, T. F., & Marx, M. B. Critical life events and coronary disease. In W. D. Gentry & R. B. Williams, Jr. (Eds.), *Psychological aspects of myocardial infarction and coronary care* (2nd ed.). St. Louis: Mosby, 1979.

Glass. D. C. *Behavior patterns, stress, and coronary disease.* New York: Wiley, 1977.

Graham, L. E., Ho, P., Thorensen, C. E., Levenkron, J. C., Vodak, P., Blair, S. N., Gelston, M., Terry, R. B., Moran, J. A., Haskell, W. L., & Wood, P. D.

Predicting treadmill fatigue suppression with various Type A behavior measures. Paper presented at the annual meeting of the American Psychological Association, Los Angeles, August 1981.

Hackett, T. P., & Cassem, H. H. Psychologic aspects of rehabilitation after myocardial infarction. In N. D. Wenger & H. K. Hellerstein (Eds.), *Rehabilitation of the coronary patient.* New York: Wiley, 1979.

Holmes, T. H., & Rahe, R. H. The social readjustment rating scale. *Journal of Psychosomatic Research,* 1967, *11,* 213–218.

Husaini, B. A., & Neff, J. A. Social class and depressive symptomatology: The role of life change events and locus of control. *Journal of Nervous and Mental Disease,* 1981, *169* (10), 638–647.

Johnson, J. H., & Sarason, I. G. Life stress, depression and anxiety: Internal-external control as a moderator variable. *Journal of Psychosomatic Research,* 1978, *22,* 205–208.

Kahn, J. P., Kornfeld, D. S., Frank, K. A., Heller, S. S., & Hoar, P. F. Type A behavior and blood pressure during coronary artery bypass surgery. *Psychosomatic Medicine,* 1980, *42* (4), 407–414.

Kornitzer, M., Magotteau, V., Degre, C., Kittel, F., Struyven, J., & vanThiel, E. Angiographic findings and the Type A pattern assessed by means of the Bortner Scale. *Journal of Behavioral Medicine,* 1982, *5* (3), 313–320.

Krantz, D. S., Glass, D. C., Schaeffer, M. A., & Davia, J. E. Behavior patterns and coronary disease: A critical evaluation. In J. R. Cacioppo & R. E. Petty (Eds.), *Perspectives in cardiovascular psychophysiology.* New York: Guilford Press, 1981.

Lau, R. R. Origins of health locus of control beliefs. *Journal of Personality and Social Psychology,* 1982, *42* (2), 322–334.

Lau, R. R., & Ware, J. F. Refinements in the measurement of health-specific locus-of-control beliefs. *Medical Care,* 1981, *19* (11), 1147–1157.

Lazarus, R. S. Psychological stress and coping in adaptation and illness. *International Journal of Psychiatry in Medicine,* 1974, *5,* 321–333.

Lorber, J. Good patients and problem patients: Conformity and deviance in a general hospital. *Journal of Health and Social Behavior,* 1975, 213–225.

Lowery, B., Jacobsen, B., & Keane, A. Relationship of locus of control to pre-operative anxiety. *Psychological Reports,* 1975, *37,* 1115–1121.

Merwin, S. L., & Abram, H. S. Psychologic response to coronary artery bypass. *Southern Medical Journal,* 1977, *70* (2), 153–155.

O'Leary, M. R., Donovan, D. M., Cysewski, B., & Chaney, E. F. Perceived locus of control, experienced control, and depression: A trait description of the learned helplessness model of depression. *Journal of Clinical Psychology,* 1977, *33* (1), 164–168.

Pilowsky, I., Spence, N. D., & Waddy, J. F. Illness behaviour and coronary artery by-pass surgery. *Journal of Psychosomatic Research,* 1979, *23,* 39–44.

Prociuk, T. J., Breen, L. J., & Lussier, R. J. Hopelessness, internal–external locus of control, and depression. *Journal of Clinical Psychology,* 1976, *32* (2), 299–300.

Rabkin, J. G., & Struening, E. L. Life events, stress, and illness. *Science,* 1976, *194,* 1013–1020.

Radley, A. R. Theory and data in the study of 'coronary proneness' (Type A behaviour pattern). *Social Science and Medicine,* 1982, *16,* 107–114.

Rahe, R. H., & Arthur, R. J. Life change patterns surrounding illness experience. *Journal of Psychosomatic Research,* 1968, *11,* 341–345.

Ramshaw, J. E., & Stanley, G. Individual differences in life-style response to coronary artery bypass surgery. *British Journal of Medical Psychology,* 1981, *54,* 83–89.

Roskies, E. Evaluating improvement in the coronary-prone (Type A) behavior pattern. In D. J. Oborne, M. M. Gruneberg, & J. R. Elser (Eds.), *Research in psychology and medicine* (Vol. 1). New York: Academic Press, 1979.

Roskies, E. Type A intervention: Finding the disease to fit the cures. In R. S. Surwit, R. B. Williams, A. Steptoe, & R. Biersner (Eds.), *Behavioral treatment of disease.* New York: Plenum, 1982.

Rotter, J. B. Generalized expectancies for internal versus external control of reinforcement. *Psychological Monographs,* 1966, *80* (1 Whole No. 609).

Rotter, J. B. Personal communication, June 1983.

Seligman, M. E. P. Learned helplessness. *Annual Revue of Medicine,* 1972, *23,* 407–412.

Selye, H. *Stress without distress.* Philadelphia: Lippincott, 1974.

Seeman, M., & Evans, J. Alienation and learning in a hospital setting. *American Sociological Review,* 1962, *127,* 772–783.

Smith, R. E. Changes in locus of control as a function of life crisis resolution. *Journal of Abnormal Psychology,* 1970, *75* (3), 328–332.

Snow, B. R., & Glass, D. C. *Differential reactivity of Type A and B individuals to congruent and incongruent environments.* Paper presentation at the 52nd annual meeting of the Eastern Psychological Association, New York, 1981. (a)

Snow, B. R., & Glass, D. C. *Coping responses of Type A and B individuals to differential performance demands.* Paper presentation at the 89th annual convention of the American Psychological Association, Los Angeles, 1981. (b)

Sparacino, J. The Type A (coronary-prone) behavior pattern, aging, and mortality. *Journal of the American Geriatrics Society,* 1979, *27* (6), 251–257.

Stensrud, R., Gilgen, A. R., & Koloc, A. Type A behavior and eastern versus western belief systems of a U.S.A. sample. *Psychological Reports,* 1982, *50,* 445–446.

Steptoe, A. *Psychological factors in cardiovascular disorders.* New York: Academic Press, 1981.

Strube, M. J., & Werner, C. *Heart rate control and the Type A coronary-prone behavior pattern.* Paper presentation at the annual meeting of the American Psychological Association, Los Angeles, August 1981.

Suinn, R. M. Intervention with Type A behaviors. *Journal of Consulting and Clinical Psychology,* 1982, *50* (6), 933–949.

Taylor, S. E. Hospital patient behavior: Reactance, helplessness, or control? *Journal of Social Issues,* 1979, *35,* 156–184.

Thurer, S., Levine, F., & Thurer, R. The psychodynamic impact of coronary bypass surgery. *International Journal Psychiatry in Medicine,* 1980–81, *10* (3), 273–290.

Vickers, R. R., Hervig, L. K., Rahe, R. H., & Rosenman, R. H. Type A behavior pattern and coping and defense. *Psychosomatic Medicine,* 1981, *43* (5), 381–396.

Weiss, J. M. Psychological factors in stress and disease. *Scientific American,* 1972, *226* (6), 103–113.

Weiss, J. M., Bailey, W. H., Goodman, P. A., Hoffman, L. J., Ambrose, M. J., Salman, S., & Charry, J. M. A model for neurochemical study of depression. In N.Y. Spiegelstein & A. Levy (Eds.), *Behavioral models and the analysis of drug action. Proceedings of the 27th OHOLO Conference,* Zichron Ya'acov, Israel, 28–31, March, 1982. Amsterdam: Elsevier Scientific Publishing Company, 1982.

Williams, R. B., Lane, J. D., Kuhn, C. M., Melosh, W., White, A. D., & Schanberg, S. M. Type A behavior and elevated physiological and neuroendocrine responses to cognitive tasks. *Science,* 1982, *218,* 483–485.

Wilson-Barnett, J. Assessment of recovery: With special reference to a study with post-operative cardiac patients. *Journal of Advanced Nursing,* 1981, *6,* 435–445.

Zyzanski, S. J., Stanton, B. A., Jenkins, C. D., & Klein, M. D. Medical and psychosocial outcomes in survivors of major heart surgery. *Journal of Psychosomatic Research,* 1981, *23* (3), 213–221.

10

Conclusions and Recommendations

JUNE B. PIMM

In this book we have attempted to describe the way in which crisis intervention can help coronary bypass patients avoid significant feelings of depression following surgery. We have included information regarding the operation itself, a description of how to do crisis intervention, and the results of research on 104 male coronary bypass patients. We have also added information on the Millon Behavioral Health Inventory, a new measure of psychological characteristics of medical populations. Ways for professionals who work with coronary bypass patients to use this scale have been provided. Chapter 3, written by a nurse, describes a new role for nurses of bypass patients, that of the patient-educator.

As a result of our research we are persuaded that crisis intervention is a helpful technique with this patient population. We have not been able to assess its effectiveness on a diverse sample of patients, however, and this should probably be done at some point. It is possible that our patient population contained some characteristics which made it particularly appropriate for this type of intervention technique. For example, there may be something psychologically unique about the coronary bypass patient himself. In other words, it is possible that there is some self-selection which takes place and patients who decide to have surgery are different in personality style than those who prefer medical management of angina.

Also, our patients were, on the whole, older and more financially successful than the average. They had chosen to leave their former homes and move to Miami, either to start a new occupation or retire after a successful career. Their internal scores on the Locus of Control

JUNE B. PIMM • Pimm Consultants, 2699 S. Bayshore Drive, Miami, Florida 33133.

Scale would suggest that, on the whole, they prefer to take responsibility for their own affairs. They are unlikely to feel that others are more capable of looking after them than they are of looking after themselves. Crisis intervention is a very appropriate means of helping this type of patient. It assists the individual in recovering his own coping ability and attempts to return him to independence as quickly as possible.

Depression was not so evident in our patients as it is often found to be in groups of coronary patients. Bypass patients may be a subgroup of heart disease patients who are particularly optimistic and well motivated to recover their health. Therefore, it is not possible to say to what extent crisis intervention would help the severely depressed patient who had accepted his role as an invalid. Although group measures of psychological functioning did not appear to affect the overall outcome of crisis intervention, it still appears that psychological factors must be taken into account on an individual basis. We have suggested that the Millon Behavioral Health Inventory, the SCL-90-R, and the Recent Life Events Questionnaire may be helpful instruments for surgeons and cardiologists.

We have also suggested that when it is not possible for a medical setting to provide crisis intervention on a regular basis, the creation of a patient-educator is a possible alternative. A chapter describes this proposal and suggests that this individual can serve as a first link in identifying those who might benefit from the crisis intervention approach.

There is no question that our main finding underscores the need for attention to psychological factors in coronary bypass surgery. Our patients, on the whole, achieved an excellent medical result. In spite of that, those in the group who did not receive crisis intervention did not appear as well satisfied as those who did. In fact, some of them experienced significant depression as late as three years after surgery. If such a costly operation, with its promised outcome of a better quality of life, is be feasible for large numbers of patients, it may be necessary to provide psychological help if that promise is to be fulfilled.

Index